SpringerBriefs in Molecular Science

SpringerBriefs in Molecular Science present concise summaries of cutting-edge research and practical applications across a wide spectrum of fields centered around chemistry. Featuring compact volumes of 50 to 125 pages, the series covers a range of content from professional to academic. Typical topics might include:

- A timely report of state-of-the-art analytical techniques
- A bridge between new research results, as published in journal articles, and a contextual literature review
- A snapshot of a hot or emerging topic
- An in-depth case study
- A presentation of core concepts that students must understand in order to make independent contributions

Briefs allow authors to present their ideas and readers to absorb them with minimal time investment. Briefs will be published as part of Springer's eBook collection, with millions of users worldwide. In addition, Briefs will be available for individual print and electronic purchase. Briefs are characterized by fast, global electronic dissemination, standard publishing contracts, easy-to-use manuscript preparation and formatting guidelines, and expedited production schedules. Both solicited and unsolicited manuscripts are considered for publication in this series.

Xinfeng Zhao · Qian Li · Jing Wang · Qi Liang ·
Jia Quan

G Protein-Coupled Receptors

Immobilization and Applications in Drug
Discovery

 Springer

Xinfeng Zhao
College of Life Sciences
Northwest University
Xi'an, China

Qian Li
College of Life Sciences
Northwest University
Xi'an, China

Jing Wang
College of Life Sciences
Northwest University
Xi'an, China

Qi Liang
College of Life Sciences
Northwest University
Xi'an, China

Jia Quan
College of Life Sciences
Northwest University
Xi'an, China

ISSN 2191-5407   ISSN 2191-5415 (electronic)
SpringerBriefs in Molecular Science
ISBN 978-981-99-0077-0   ISBN 978-981-99-0078-7 (eBook)
https://doi.org/10.1007/978-981-99-0078-7

This Springer imprint is published by the registered company Springer Nature Singapore Pte Ltd.
The registered company address is: 152 Beach Road, #21-01/04 Gateway East, Singapore 189721, Singapore

Preface

G protein-coupled receptors (GPCRs) are central in mediating many physiological responses and widely accepted as the largest family of drug targets. Many techniques relying on the immobilized GPCRs have been developed for high-throughput drug-receptor interaction analysis and lead screening. However, these works have been limited by difficulties in GPCR immobilization with high activity, high stability, and conformational selectivity. Herein, we provide an overview of the structures, functions, and purification of GPCR and the progress in the oriented immobilization strategy, including affinity-based non-covalent immobilization, site-specific covalent immobilization through bioorthogonal chemistry, and biologically mediated site-specific immobilization. We also introduce the key biochemical aspects of drug-target interactions and discuss the immobilized GPCRs in the drug-receptor interaction analysis. Finally, we review the immobilized GPCRs in lead compound screening from compound libraries and natural product extracts.

Finally, we thank the financial support from the National Natural Science Foundation of China (22074118, 22004097, 21974107, 82174008, 21775119, 21705126, 21475103, and 21005060), the Key Research and Development Program of Shaanxi Province (2020ZDLSF05-07), the Innovation Capability Support Program of Shaanxi Province (2022KJXX-70), and the Shaanxi Administration of Traditional Chinese Medicine (2021-04-22-005 and 2022-SLRH-YQ-007).

Xi'an, China

Xinfeng Zhao
Qian Li
Jing Wang
Qi Liang
Jia Quan

Perspectives

Two decades of intense research and development have led to the considerable success of the technology applications that facilitate studies of GPCR-related drug discovery. At least one-third of these techniques rely on the immobilized GPCRs. An ideal immobilization strategy should involve minimal alterations of the native structure of GPCRs and maintain the high activity, high stability, and conformational selectivity of the receptors. Although the explosive development of molecular cloning makes the heterologous expression of GPCRs with desired protein tags or mutations more tractable, which enables a large number of immobilization strategies to be approached by tagging (protein labeling) system. The characterization of the immobilized receptors remains open, especially in the measurement of the numbers of receptors, the conformational change of the bound and unbound receptors, and the determination of the receptor downstream signaling pathways. What kind of new GPCR immobilization method will be developed in the next decade? Will the method improve the accuracy of screening and accelerate the drug discovery process? What new challenges and discoveries will be made? One thing is certain: the oriented immobilization-based methods have offered new insights for drug discovery that target GPCR.

Contents

About the Authors

Xinfeng Zhao is a Professor in the College of Life Sciences at Northwest University (Xi'an, China) since 2016. He received his B.S. and M.S. in Chemistry from Northwest University in 2001 and Xi'an Jiaotong University in 2005, respectively. He obtained his doctorate in pharmacology in 2009 at Xi'an Jiaotong University and then pursued postdoctoral studies in pharmacology at the University of Cambridge (Cambridge, UK) in 2010. In 2009, he joined Northwest University as a Lecturer and was promoted to Associate Professor in 2013. He has been a Senior Visiting Scholar in the Department of Chemistry at Purdue University (West Lafayette, IN, USA) in 2017. His research interests are in the areas of immobilized GPCRs and applications, GPCR-drug interaction analysis, and bioactive compound screening from herbal medicines.

Qian Li received her doctorate in Pharmaceutical Analysis in 2016 and pursued postdoctoral studies in biology at Northwest University (Xi'an, China). Her dissertation focused on the establishment and application of chromatographic methods based on immobilized GPCRs. In 2018, she joined Northwest University as a Lecturer and was appointed an Associate Professor in the College of Life Sciences in 2020. In 2018, she moved to the Department of Chemistry at Purdue University (West Lafayette, IN, USA) as a visiting scholar. Her current interests lie in the area of immobilized GPCRs and their applications.

Jing Wang joined the faculty as a Lecturer in the College of Life Sciences at Northwest University (Xi'an, China). She received her Ph.D. in Medicinal Chemistry in 2019 from Northwest University under the direction of Professor Xinfeng Zhao. Her research efforts are primarily directed toward the immobilization of GPCRs (e.g., beta2-adrenoceptor and leukotriene receptor) through bioorthogonal reaction and their application in receptor-ligand interaction analysis and lead compound screening from natural products.

Qi Liang received his B.S. in Life Sciences and M.S. in Pharmacology from Northeast Forestry University (Harbin, China) in 2010 and Xi'an Jiaotong University

(Xi'an, China) in 2013, respectively. In 2021, he obtained his Ph.D. in medicinal chemistry from Northwest University (Xi'an, China) under the direction of Prof. Xinfeng Zhao. Afterward, he was engaged in his postdoctoral work in the same group. His research interests are screening the ligands of GPCR from a DNA-encoded library.

Jia Quan received his B.S. and Ph.D. from Northwest University (Xi'an, China) in 2015 and Wuhan University (Wuhan, China) in 2021, respectively. In 2022, he joined the group of Prof. Xinfeng Zhao at Northwest University as a postdoctoral fellow. His research project focuses on the virus-encoded GPCRs and the screening for new antivirus drugs.

Chapter 1
G Protein-Coupled Receptors

Abstract G protein-coupled receptors (GPCRs) are important drug targets for FDA-approved drugs. Understanding the biology, structure, and pharmacology of GPCRs is of paramount importance for drug discovery. This chapter summarizes the structures, conformations, and signaling pathways of different GPCRs, providing an overview of GPCRs for drug discovery purposes.

Keywords G protein-coupled receptor · Structure and conformation · Signaling pathway

Abbreviations

GPCRs	G-protein coupled receptors
$5\text{-}HT_{1B}R$	5-Hydroxytryptamine 1B receptor
$5\text{-}HT_{2A}R$	5-Hydroxytryptamine 2A receptor
$5\text{-}HT_{2B}R$	5-Hydroxytryptamine 2B receptor
$5\text{-}HT_{2C}R$	5-Hydroxytryptamine 2C receptor
M_1R	Muscarinic M_1 receptor
M_2R	Muscarinic M_2 receptor
M_4R	Muscarinic M_4 receptor
M_5R	Muscarinic M_5 receptor
$\alpha_{2A}\text{-}AR$	α_{2A}-Adrenoceptor
$\alpha_{2B}\text{-}AR$	α_{2B}-Adrenoceptor
$\alpha_{2C}\text{-}AR$	α_{2C}-Adrenoceptor
$\beta_1\text{-}AR$	β_1-Adrenoceptor
$\beta_2\text{-}AR$	β_2-Adrenoceptor
D_2R	Dopamine D_2 receptor
D_3R	Dopamine D_3 receptor
D_4R	Dopamine D_4 receptor
H_1R	Histamine H_1 receptor
AT_1R	Angiotensin II type 1 receptor
AT_2R	Angiotensin II type 2 receptor

X. Zhao et al., *G Protein-Coupled Receptors*, SpringerBriefs in Molecular Science,
https://doi.org/10.1007/978-981-99-0078-7_1

APJ	Apelin receptor
$C5_{a1}R$	$C5_{a1}$ receptor
$C5_{a2}R$	$C5_{a2}$ receptor
ET_BR	Endothelin B receptor
FPR_1	Formyl peptide receptor 1
FPR_2	Formyl peptide receptor 2
$GnRH_1R$	Gonadotropin releasing hormone 1 receptor
MC_4R	Melanocortin 4 receptor
LPA_1R	Lysophosphatidic acid receptor type 1
Y_1R	Neuropeptide Y receptor type 1 receptor
NTS_1R	Neurotensin receptor 1
δ-OR	δ-Opioid receptors
κ-OR	κ-Opioid receptors
μ-OR	μ-Opioid receptors
NOPR	Nociceptin opioid peptide receptor
OX_1R	Orexin 1 receptor
OX_2R	Orexin 2 receptor
PAR1	Protease-activated receptor-1
PAR2	Protease-activated receptor-2
NK_1R	Neurokinin 1 receptor
$V_{1A}R$	Vasopressin receptor 1A
V_2R	Vasopressin receptor 2
OTR	Oxytocin receptor
$ACKR_1$	Atypical chemokine receptor 1
FFA_1R	Free fatty acid receptor 1
$CysLTR_1$	Cysteinyl leukotriene receptor 1
$CysLTR_2$	Cysteinyl leukotriene receptor 2
$LPAR_1$	Lysophosphatidic acid receptor 1
S_1P_1R	Sphingosine-1-phosphate receptor 1
CB_1R	Cannabinoid receptor 1
PAFR	Platelet-activating factor receptor
DP_1R	D-type prostanoid receptor 1
EP_2R	Prostaglandin E2 receptor
EP_3R	Prostaglandin E2 receptor subtype 3
EP_4R	Prostaglandin E2 receptor subtype 4
TPR	Thromboxane receptor
A_1R	Adenosine A_1 receptor
$A_{2A}R$	Adenosine A_{2A} receptor
P_2Y_1R	Purinergic P_2Y_1 receptor
$P_2Y_{12}R$	Purinergic P_2Y_{12} receptor
GPBAR	Bile acid receptor
GPR52	G-protein coupled receptor 52
MT_1R	Melatonin receptor type 1A
MT_2R	Melatonin receptor 1B
Ghs-R	Growth hormone secretagogue receptor

CCR$_2$	C-C chemokine receptor type 2
CCR$_5$	C-C chemokine receptor type 5
CCR$_6$	C-C chemokine receptor type 6
CCR$_7$	C-C chemokine receptor type 7
CCR$_9$	C-C chemokine receptor type 9
CXCR$_2$	C-X-C motif chemokine receptor 2
CXCR$_4$	C-X-C motif chemokine receptor 4
FZD$_4$	Frizzled class receptor 4
FZD$_5$	Frizzled class receptor 5
SMO	Smoothened receptor
CTR	Calcitonin receptor
CTLR	Calcitonin receptor-like receptor
CRF$_1$	Corticotropin-releasing factor type 1 receptor
CRF$_2$	Corticotropin-releasing factor type 2 receptor
GHRHR	Growth-hormone-releasing hormone receptor
GLP$_1$R	Glucagon-like peptide-1 receptor
GLP$_2$R	Glucagon-like peptide-2 receptor
GCGR	Glucagon receptor
SR	Secretin receptor
PTH$_1$R	Parathyroid hormone type 1 receptor
PAC$_1$R	Pituitary adenylate cyclase-activating polypeptide receptor type 1 receptor
VPAC$_1$R	Vasoactive intestinal polypeptide receptor 1

1.1 GPCR Biology

As the largest human transmembrane protein superfamily, G protein-coupled receptors (GPCRs) have more than 800 members, accounting for 35% of FDA-approved drugs (Calebiro et al. 2021; Cornwell and Feigin 2020; Hauser et al. 2017). These receptors mediate almost every major therapeutic category or disease class, including cardiovascular, metabolic, and nervous system ailments (Pfleger et al. 2019; Rosenbaum et al. 2020). This makes GPCRs the most important target among the 800 reported drug targets. The development of new strategies for drug discovery based on GPCRs has become the central part of the whole drug discovery process.

Immobilization of target proteins onto a solid surface is fundamental to the successful development of a series of techniques that have played a particularly important role in clinical diagnosis and drug discovery (Zou et al. 2018). Such a strategy has been limited to GPCRs ascribed to two major obstacles (Früh et al. 2011). First, obtaining accurate crystal structures of GPCRs is a huge challenge due to the resemblance between different GPCRs and their diverse conformations. Second, GPCRs are seven-transmembrane proteins with poor solubility and structural changes and the loss of bioactivity could happen during the immobilization of the receptors. The breakthroughs in X-ray crystallography for identifying three-dimensional structures of GPCRs have led to the rapid progress of immobilized GPCRs in diverse areas (Kang et al. 2015). For instance, new strategies with high specificity for oriented immobilization of GPCRs, immobilized GPCRs for drug-receptor analysis, and application of immobilized GPCRs in diverse stages of the drug discovery process (Smith et al. 2013). These researches have provided proof of broad prospects for immobilized GPCRs in most life science-associated areas.

In recent reports, the amazing superiority of immobilized GPCRs is also confirmed, especially in high throughput drug-receptor interaction analysis and lead screening (Feng et al. 2020). In this case, a number of new techniques including automated radiometric immunoassays, proximity chemiluminescence assays, new surface plasmon resonance, and impedance biosensor-based assays and affinity chromatography have been successfully established depending on immobilized GPCRs. Application of these assays in different stages of the drug discovery process has also been made and has resulted in several clinical drugs. These new advances necessitate the comprehensive introduction of assays for GPCR immobilization and the application of these assays in drug discovery to chemical and pharmaceutical areas.

This review intends to comprehensively summarize new strategies for oriented immobilization of GPCRs and their new applications in diverse stages of the drug discovery process. We will also outline emerging and future lines of inquiry for the fabrication of immobilized GPCR assays and harnessing them for drug discovery.

1.2   GPCR Structures

The recent completion of the human genome sequence allows the identification of about 800 GPCRs that should be included in the IUPHAR receptor classification. More than half of these GPCRs have sensory functions, mediating olfaction (~400), taste (33), light perception (10), and pheromone signaling (5) (Mombaerts 2004). The remaining non-sensory GPCRs (~350) are canonical targets for the majority of drugs in clinical practice (Chen et al. 2012; Overington et al. 2006; Russ and Lampel 2005). They mediate the signaling by ligands sized from small molecules to peptides to large proteins. The characteristic feature of these GPCRs is the presence of seven α-helical transmembrane domains. There are also extensive amino acid sequence similarities that divide them into several classes, each with characteristic highly conserved residues distributed throughout the molecule, which define identifying motifs, such as the DRY motif at the cytoplasmic end of the third transmembrane domain and prolines at specific positions in helices 5, 6, and 7 of the very large class related to rhodopsin (Rovati et al. 2007). The first classification scheme by Kolakowski et al. has divided GPCRs into six classes relying on sequence homology (Kolakowski 1994). These classes and their prototype members included: Class A (rhodopsin-like), Class B (secretin receptor family), Class C (metabotropic glutamate), Class D (fungal mating pheromone receptors), Class E (cyclic AMP receptors), and Class F (frizzled/smoothened). Of these, classes D and E are not found in vertebrates.

An alternative classification scheme "GRAFS" divides vertebrate GPCRs into five classes, overlapping with the A-F nomenclature, viz. Rhodopsin family (class A), Glutamate family (class C), Adhesion family, Frizzled family, and Secretin family (Krishnan et al. 2014).

Prior to 2007, GPCR structure has been extensively probed using biophysical and biochemical methods coupled with interpretation via ab initio or rhodopsin-based modeling (Schwartz et al. 2006; Kobilka 2007). In this period, structural information on the highly diverse repertoire of GPCRs has been limited to classes A, B, and C with the prototypes of rhodopsin (Kang et al. 2015), secretin receptor (Dong et al. 2020), and metabotropic glutamate receptors (Christopher et al. 2019), respectively (Bockaert and Pin 1999; Josefsson 1999; Graul and Sadée 2001; Joost and Methner 2002; Fredriksson et al. 2003). By June 2021, advances in protein engineering and crystallography have galvanized exponential growth in the structure determination of human GPCRs. Eighty-six diverse GPCRs had been determined and have been traditionally considered canonical targets in pharmacology and drug development (Table 1.1). The receptors from α-group Class A GPCRs include rhodopsin, four 5-hydroxytryptamine receptors (5-HT$_{1B}$R (Wang et al. 2013), 5-HT$_{2A}$R (Kimura et al. 2019), 5-HT$_{2B}$R (Wacker et al. 2013), and 5-HT$_{2C}$R (Peng et al. 2018), four muscarinic acetylcholine receptors M$_1$R, M$_2$R, M$_4$R, and M$_5$R (Maeda et al. 2020; Suno et al. 2018; Thal et al. 2016; Vuckovic et al. 2019; Xu et al. 2019; Haga et al. 2012; van Koppen and Kaiser 2003), five adrenergic receptors α_{2A}-AR, α_{2B}-AR, α_{2C}-AR, β_1-AR, and β_2-AR (Yuan et al. 2020; Rasmussen et al. 2007; Ma et al. 2017b; Cherezov et al. 2007; Warne et al. 2008), three dopamine receptors D$_2$R (Fan et al. 2020), D$_3$R (Chien et al. 2010), and D$_4$R (Wang et al. 2017), histamine H1 receptor (H$_1$R) (Shimamura et al. 2011), melanocortin 4 receptor (MC$_4$R) (Yu et al. 2020), lysophosphatidic acid receptor type 1 (LPA$_1$R) (Chrencik et al. 2015), cannabinoid receptor 1 (CB$_1$R) (Li et al. 2019), prostaglandin receptors (DP$_1$R (Wang et al. 2018), EP$_2$R (Qu et al. 2021), EP$_3$R (Morimoto et al. 2019), and EP$_4$R (Toyoda et al. 2019), thromboxane receptor (TPR) (Fan et al. 2019), two adenosine receptors A$_1$R and A$_{2A}$R (Glukhova et al. 2017; Lebon et al. 2011; Jaakola et al. 2008), G-protein coupled receptor 52 (GPR52) (Lin et al. 2020), two melatonin receptors (MT$_1$R and MT$_2$R) (Stauch et al. 2019; Johansson et al. 2019) and the first example of a lipid-activated GPCR, sphingosine-1-phosphate receptor 1 (S$_1$P$_1$) (Hanson et al. 2012). A number of peptide-binding receptors from the β group of Class A GPCRs have also been characterized such as the endothelin B receptor (ET$_B$R) (Nagiri et al. 2019), gonadotropin-releasing hormone 1 receptor (GnRH$_1$R) (Yan et al. 2020), neuropeptide Y receptor type 1 receptor (Y$_1$R) (Yang et al. 2018), neurotensin receptor 1 (NTS$_1$R) (Huang et al. 2020), two orexin receptors (OX$_1$R and OX$_2$R) (Yin et al. 2015, 2016), neurokinin 1 receptor (NK$_1$R) (Schöppe et al. 2019). wo vasopressin receptors (V$_{1A}$R and V$_2$R) (Adikesavan et al. 2005; Staus et al. 2020), oxytocin receptor (OTR) (Waltenspühl et al. 2020), and growth hormone secreta-gogue receptor (Ghs-R) (Shiimura et al. 2020). The γ group of Class A GPCRs contain two angiotensin II receptors (AT$_1$R (Zhang et al. 2015a, b) and AT$_2$R (Asada et al. 2020), two C5$_a$ receptors (C5$_{a1}$R (Robertson et al. 2018) and C5$_{a2}$R (Colley et al. 2018)), formyl peptide receptors (FPR$_1$ and FPR$_2$) (Douthwaite et al. 2015;

Chen et al. 2020), the chemokine receptors $CXCR_2$ (Liu et al. 2020) and $CXCR_4$ (Wu et al. 2010), atypical chemokine receptor 1 ($ACKR_1$) (Batchelor et al. 2014), C–C chemokine receptors CCR_2 (Apel et al. 2019), CCR_5 (Tan et al. 2013), CCR_6 (Wasilko et al. 2020), CCR7 (Jaeger et al. 2019), and CCR_9 (Oswald et al. 2016), the κ-opioid receptor (κ-OR) (Che et al. 2017), the μ-OR (Koehl et al. 2018), the δ-OR (Fenalti et al. 2014), and the nociceptin opioid peptide receptor (NOP) (Miller et al. 2015). Beyond these Class A GPCRs, the structures of the apelin receptor (APJ) (Ma et al. 2017a, b), two protease-activated receptors (PAR1 (Zhang et al. 2012) and PAR2 (Cheng et al. 2017), two cysteinyl leukotriene receptors ($CysLTR_1$ (Luginina et al. 2019) and $CysLTR_2$ (Gusach et al. 2019)), platelet-activating factor receptor (PAFR) (Cao et al. 2018), and two purinergic receptors (P_2Y_1R (Zhang et al. 2015a, b) and $P_2Y_{12}R$ (Zhang et al. 2014)) from δ-group were also resolved.

Among the deposited GPCRs, opioid receptors are the most thoroughly discussed subfamily. The crystal structures of the four closely related opioid subtypes have been solved followed by adrenergic receptors, muscarinic acetylcholine receptors, 5-hydroxytryptamine receptors, C–C chemokine receptors, dopamine receptors, and prostaglandin receptors. Most of the solved structures are single representatives of their subfamilies. As listed in Table 1.1, a considerable structural determination has been achieved for the Class A GPCRs, including 32 α-group receptors, 11 β-group receptors, 18 γ-group receptors, and 8 δ-group receptors. In contrast to Class A GPCRs, the structural coverages of other families of GPCRs thus far remain relatively sparse.

1.3   GPCR Pharmacology

1.3.1   G Protein-Coupled Receptor Conformations

The crystal structures of several GPCRs, such as rhodopsin (Kang et al. 2015), $β_2$-AR (Rasmussen et al. 2007), $β_1$-AR (Emtage et al. 2017), A_{2A}-R (Lebon et al. 2011; Xu et al. 2011) and $CXCR_4$ (Qin et al. 2015) have been accomplished by co-crystalizing the receptor in complexes with diverse ligands, in distinct crystal forms, or by different methodologies for crystallization (Table 1.2). Despite a similar seven transmembrane (7TM) topology, the five major human GPCR families display little sequence identity and possess distinct extracellular N-terminal domains. Such structural variations in extracellular loops, TM helices, and side chains create a remarkable variety of sizes, shapes, and electrostatic properties of the ligand-binding pockets in different GPCR subfamilies, resulting in the diversity of their conformations for corresponding ligands.

Inspired by crystallographic characterization of rhodopsin, A_{2A}-R, and $β_2$-AR (Fig. 1.1), GPCRs have appeared to have five conformationally distinct functional states due to a dynamic equilibrium between inactive (R, R′) and active (R″, R*) states. Further transformation of the active states into the signaling state (R * G)

Table 1.1 Crystal structures of human GPCRs

Receptor	Gene name	No	PDB code(s)	Year(s)	Resolution (Å)	Family
5-HT$_{1B}$R	HTR1B	5	4IAR, 4IAQ, 5V54, 7C61, 6G79	2013–2020	2.7–3.9	α
5-HT$_{2A}$R	HTR2A	4	6A94, 6A93, 6WGT, 6WH4	2019–2020	2.9–3.4	α
5-HT$_{2B}$R	HTR2B	8	4IB4, 5TUD, 4NC3, 6DRX, 5TVN, 6DRY, 6DRZ, 6DS0	2013–2018	2.7–3.188	α
5-HT$_{2C}$R	HTR2C	2	6BQG, 6BQH	2018	2.7–3.0	α
M$_1$R	CHRM1	3	6WJC, 5CXV, 6OIJ	2020–2016	2.55–3.3	α
M$_2$R	CHRM2	10	5ZKC, 5ZKB, 5ZK8, 5ZK3, 5YC8, 3UON, 4MQS, 4MQT, 6OIK, 6U1N	2012–2020	2.3–4.0	α
M$_4$R	CHRM4	2	5DSG, 6KP6	2013–2020	2.6–3	α
M$_5$R	CHRM5	1	6OL9	2019	2.541	α
α$_{2A}$-AR	ADRA2A	2	6KUY, 6KUX	2019	2.7–3.2	α
α$_{2B}$-AR	ADRA2B	2	6K41, 6K42	2020	2.9–4.1	α
α$_{2C}$-AR	ADRA2C	1	6KUW	2019	2.8	α
β$_1$-AR	ADRB1	4	7BU6, 7BU7, 7BTS, 6TKO	2020	2.6–3.3	α
β$_2$-AR	ADRB2	38	2R4R, 2RH1, 3D4S, 2R4S, 3KJ6, 3NY8, 3NY9, 3NYA, 3PDS, 4LDL, 4LDO, 4LDE, 3P0G, 4QKX, 6E67, 6PS0, 5D5A, 5D5B, 5D6L, 5JQH, 5X7D, 4GBR, 6PS6, 6PS5, 6PRZ, 6PS2, 6PS1, 6PS4, 6PS3, 3SN6, 6OBA, 5X7D, 6NI3, 6MXT, 6N48, 7DHR, 7DHI, 7BZ2	2007–2020	2.703–3.993	α
Rhodopsin	RHO	7	4ZWJ, 5DGY, 5W0P, 6FUF, 6CMO, 6QNO, 5AFP	2015–2019	2.30–7.70	α
D$_2$R	DRD2	6	6LUQ, 6CM4, 7DFP, 6VMS, 5AER, 7JVR	2015–2021	2.19–3.1	α
D$_3$R	DRD3	3	3PBL, 7CMU, 7CMV	2010–2021	2.89–3.0	α
D$_4$R	DRD4	2	5WIV, 5WIU	2017	1.962–2.143	α
H$_1$R	HRH1	2	3RZE, 7DFL	2011–2021	3.1–3.3	α

(continued)

Table 1.1 (continued)

Receptor	Gene name	No	PDB code(s)	Year(s)	Resolution (Å)	Family
MC4R	MC4R	2	6W25, 7AUE	2020–2021	2.75–2.97	α
LPA1R	LPAR1	3	4Z34, 4Z35, 4Z36	2015	2.9–3	α
CB1R	CNR2	4	5ZTY, 6KPC, 6PT0, 6KPF	2019–2020	2.8–3.2	α
DP1R	PTGDR2	2	6D26, 6D27	2018	2.738–2.798	α
EP2R	PTGER2	3	7CX2, 7CX3, 7CX4	2021	2.8–2.9	α
EP3R	PTGER3	2	6AK3, 6M9T	2018	2.5–2.9	α
EP4R	PTGER4	3	5YWY, 5YHL, 7D7M	2018–2020	3.2–4.2	α
TPR	TBXA2R	2	6IIU, 6IIV	2018	2.5–3	α
A1R	ADORA1	3	5UEN, 6D9H, 5N2S	2017–2018	3.2–3.6	α
A2AR	ADORA2A	58	2YDO, 2YDV, 3QAK, 5WF5, 5WF6, 3EML, 3REY, 3RFM, 3PWH, 3VGA, 3VG9, 3UZA, 3UZC, 4EIY, 4UG2, 4UHR, 5VRA, 5NM2, 5NM4, 5G53, 6S0L, 5UVI, 5K2D, 5K2A, 5K2C, 5K2B, 5N2R, 5OLH, 5OLZ, 5OLV, 5OM1, 5OM4, 5IU4, 5IU8, 5IU7, 5IUB, 5IUA, 5NLX, 5OLO, 5MZP, 5JTB, 5UIG, 5OLG, 5MZJ, 6PS7, 6LPK, 6LPI, 6LPL, 6S0Q, 6GDG, 6ZDR, 6ZDV, 6WQA, 6AQF, 6GT3, 6MH8, 6JZH, 7ARO	2011–2020	1.7–4.2	α
GPR52	GPR52	4	6LI0, 6LI1, 6LI2, 6LI3	2020	2.2–3.3	α
MT1R	MTNR1A	5	6ME2, 6ME3, 6ME4, 6ME5, 6PS8	2019	2.8–3.3	α
MT2R	MTNR1B	4	6ME6, 6ME7, 6ME8, 6ME9	2019	2.8–3.3	α
S1P1R	S1PR1	2	3V2W, 3V2Y	2012	2.8–3.35	β
ETBR	EDNRB	8	6K1Q, 6IGL, 6IGK, 5GLI, 5GLH, 6LRY, 5X93, 5XPR	2016–2020	2–3.6	β
GnRH1R	GNRHR	1	7BR3	2020	2.79	β
Y1R	NPY1R	2	5ZBQ, 5ZBH	2018	2.7–3	β
NTS1R	NTSR1	8	6UP7, 6OSA, 6OS9, 6PWC, 7L0S, 7L0Q, 7L0R, 7L0P	2019–2021	3.0–4.9	β

(continued)

Table 1.1 (continued)

Receptor	Gene name	No	PDB code(s)	Year(s)	Resolution (Å)	Family
OX$_1$R	HCRTR1	14	4ZJC, 4ZJ8, 6TP3, 6TOT, 6TP4, 6TP6, 6TQ4, 6TQ6, 6TQ7, 6TQ9, 6V9S, 6TOD, 6TO7, 6TOS	2016–2020	2.11–3.5	β
OX$_2$R	HCRTR2	7	4S0V, 5WQC, 5WS3, 6TPN, 6TPG, 7L1V, 7L1U	2015–2021	1.96–3.2	β
NK$_1$R	TACR1	6	6HLL, 6J21, 6J20, 6E59, 6HLO, 6HLP	2018–2019	2.2–3.27	β
V$_{1A}$R	AVPR1A	1	1YTV	2005	1.8	β
V$_2$R	AVPR2	2	6U1N, 6NI2	2020	3.8	β
OTR	OXTR	1	6TPK	2020	3.2	β
Ghs-R	GHSR	1	6KO5	2020	3.3	β
AT$_1$R	AGTR1	6	4ZUD, 4YAY, 6DO1, 6OS2, 6OS1, 6OS0	2015–2020	2.79–2.901	γ
AT$_2$R	AGTR2	6	6JOD, 5UNG, 5UNF, 5UNH, 5XJM, 7C6A	2017–2020	2.8–3.4	γ
C5$_{a1}$R	C5AR1	3	5O9H, 6C1Q, 6C1R	2015–2018	2.12–2.9	γ
C5$_{a2}$R	C5AR2	1	4UU9	2015	2.12	γ
FPR$_1$	FPR1	1	4UV4	2014	3.08	γ
FPR$_2$	FPR2	2	6LW5, 6OMM	2020	2.8–3.17	γ
δ-OR	OPRD1	5	4N6H, 6PT3, 6PT2, 4RWA, 4RWD	2012–2019	1.8–3.3	γ
κ-OR	OPRK1	2	6B73, 6VI4	2018–2020	3.1–3.3	γ
μ-OR	OPRM1	2	6DDF, 6DDE	2018	3.5–3.8	γ
NOPR	OPRL1	3	5DHH, 5DHG, 4EA3	2012–2015	3–3.013	γ
ACKR$_1$	ACKR1	2	4NUU, 4NUV	2014	1.95–2.6	γ
CCR$_2$	CCR2	3	6GPX, 6GPS, 5T1A	2016–2019	2.7–3.3	γ
CCR$_5$	CCR5	6	4MBS, 6AKY, 6AKX, 5UIW, 6MEO, 6MET	2013–2018	2.204–2.8	γ

(continued)

Table 1.1 (continued)

Receptor	Gene name	No	PDB code(s)	Year(s)	Resolution (Å)	Family
CCR_6	CCR6	1	6WWZ	2020	3.34	γ
CCR_7	CCR7	1	6QZH	2019	2.1	γ
CCR_9	CCR9	1	5LWE	2016	2.8	γ
$CXCR_2$	CXCR2	3	6LFL, 6KVF, 6KVA	2020	2.2–3.2	γ
$CXCR_4$	CXCR4	6	3OE9, 3ODU, 3OE8, 3OE0, 3OE6, 4RWS	2010–2015	2.5–3.2	γ
APJ	APLNR	2	5VBL, 6KNM	2017–2020	2.6–3.2	δ
PAR1	F2R	1	3VW7	2012	2.2	δ
PAR2	F2RL1	3	5NDD, 5NDZ, 5NJ6	2017	2.801–4	δ
$CysLT_1R$	CYSLTR1	2	6RZ5, 6RZ4	2019	2.53–2.7	δ
$CysLT_2R$	CYSLTR2	4	6RZ6, 6RZ7, 6RZ8, 6RZ9	2019	2.43–2.73	δ
PAFR	PTAFR	2	5ZKQ, 5ZKP	2018	2.81–2.9	δ
P_2Y_1R	P2RY1	2	4XNV, 4XNW	2015	2.2–2.7	δ
$P_2Y_{12}R$	P2RY12	3	4PXZ, 4PY0, 4NTJ	2014	2.5–3.1	δ
FZD_4	FZD4	3	5CL1, 6BD4, 6NE1	2015–2019	2.4–3.8	F
FZD_5	FZD5	4	6WW2, 6O39, 5URY, 5URZ	2017–2020	1.8–3.7	F
SMO	SMO	14	4JKV, 4N4W, 5V57, 5V56, 5L7I, 5L7D, 6XBK, 6XBJ, 6XBM, 6XBL, 4QIM, 4O9R, 4QIN, 6OT0	2013–2020	2.9–3.96	F
CTR	CALCR	4	6PFO, 6PGQ, 6NIY, 5UZ7	2017–2020	1.78–4.1	B1
CTLR	CALCRL	9	6ZHO, 6ZIS, 6V2E, 4RWF, 4RWG, 5V6Y, 6UMG, 7KNT, 7KNU	2015–2021	1.76–3.49	B1
CRF_1R	CRFR1	7	4K5Y, 4Z9G, 6PB0, 6P9X, 3EHS, 3EHT, 3EHU	2008–2020	1.96–3.4	B1
CRF_2R	CRFR2	1	6PB1	2020	2.8	B1

(continued)

Table 1.1 (continued)

Receptor	Gene name	No	PDB code(s)	Year(s)	Resolution (Å)	Family
GHRHR	GHRHR	1	7CZ5	2020	2.6	B1
GLP$_1$R	GLP1R	17	6ORV, 6B3J, 6KK1, 5VA1, 5VEW, 5VEX, 5NX2, 6KJV, 6KK7, 6VCB, 7LCJ, 7LCK, 7C2E, 6XOX, 6X1A, 6X18, 6X19	2017–2020	2.1–4.2	B1
GLP$_2$R	GLP2R	1	7D68	2020	3	B1
GCGR	GCGR	13	6WHC, 4L6R, 5EE7, 5XEZ, 5XF1, 5YQZ, 6WPW, 5VEW, 5VEX, 5NX2, 6LML, 6LMK, 6B3J	2013–2020	2.7–3.3.7	B1
SR	SCTR	3	7D3S, 6WZG, 6WI9	2020	2.3–4.3	B1
PTH$_1$R	PTH1R	4	6NBF, 6NBI, 6NBH, 6FJ3	2018–2019	2.5-4	B1
PAC$_1$R	ADCYAP1R1	2	6P9Y, 6LPB	2020	3.01–3.9	B1
VPAC$_1$R	VIPR1	1	6VN7	2020	3.2	B1
GPBAR	GPBAR1	3	7BW0, 7CFM, 7CFN	2020	3–3.9	Other
FFA$_1$R	FFAR1	4	4PHU, 5TZY, 5TZR, 5KW2	2014–2018	2.2–3.22	Other

Note Data were collected from protein data bank https://www.rcsb.org/

Table 1.2 Five distinct activation states represented by crystal structures of human GPCRs

Receptor	R (inactive state)	R′ (inactive agonist-bound state)	R″ (active state)	R* (active state, with G protein subunit)	R*G (G protein signaling state)
5-HT$_{1B}$R	5V54	4IAR, 4IAQ	N/A	6G79	7C61
5-HT$_{2A}$R	6A94, 6A93	N/A	N/A	6WGT, 6WH4	N/A
5-HT$_{2B}$R	N/A	4IB4, 5TVN	6DRX, 6DRY, 6DRZ, 6DS0	N/A	5TUD, 4NC3
5-HT$_{2C}$R	6BQH	6BQG	N/A	N/A	N/A
M$_1$R	6WJC, 5CXV	N/A	N/A	N/A	6OIJ
M$_2$R	5ZKC, 5ZKB, 5ZK8, 5ZK3, 5YC8, 3UON	N/A	N/A	N/A	4MQS, 4MQT, 6OIK, 6U1N
M$_4$R	5DSG	6KP6	N/A	N/A	N/A
M$_5$R	6OL9	N/A	N/A	N/A	N/A
α$_{2A}$-AR	N/A	N/A	N/A	N/A	6KUY, 6KUX
α$_{2B}$-AR	N/A	N/A	N/A	6K41, 6K42	N/A
α$_{2C}$-AR	N/A	N/A	N/A	N/A	6KUW
β$_1$-AR	N/A	N/A	N/A	7BU6, 7BU7, 7BTS	6TKO
D$_2$R	6LUQ, 6CM4, 7DFP	N/A	N/A	N/A	6VMS, 5AER, 7JVR
D$_3$R	3PBL	N/A	N/A	7CMU, 7CMV	N/A
D$_4$R	N/A	N/A	N/A	N/A	5WIV, 5WIU
H$_1$R	3RZE	N/A	N/A	7DFL	N/A

(continued)

Table 1.2 (continued)

Receptor	R (inactive state)	R' (inactive agonist-bound state)	R'' (active state)	R* (active state, with G protein subunit)	R*G (G protein signaling state)
AT_1R	4ZUD, 4YAY	N/A	6DO1, 6OS2, 6OS1, 6OS0	N/A	N/A
AT_2R	N/A	N/A	N/A	N/A	6JOD, 5UNG, 5UNF, 5UNH, 5XJM, 7C6A
APJ	N/A	5VBL	6KNM	N/A	N/A
$C5_{a1}R$	5O9H, 6C1Q, 6C1R	N/A	N/A	N/A	N/A
$C5_{a2}R$	4UU9	N/A	N/A	N/A	N/A
ET_BR	5X93, 5XPR	6K1Q, 6LRY	N/A	N/A	6IGL, 6IGK, 5GLI, 5GLH
FPR_1	4UV4	N/A	N/A	N/A	N/A
FPR_2	N/A	N/A	6LW5	N/A	6OMM
$GnRH_1R$	N/A	N/A	N/A	N/A	7BR3

(continued)

Table 1.2 (continued)

Receptor	R (inactive state)	R' (inactive agonist-bound state)	R'' (active state)	R* (active state, with G protein subunit)	R*G (G protein signaling state)
MC$_4$R	6W25	N/A	N/A	7AUE	N/A
Y$_1$R	N/A	N/A	N/A	N/A	5ZBQ, 5ZBH
NTS$_1$R	6UP7	N/A	N/A	N/A	6OSA, 6OS9, 6PWC, 7L0S, 7L0Q, 7L0R, 7L0P
δ-OR	N/A	4RWA, 4RWD	6PT3, 6PT2	N/A	4N6H
κ-OR	4DJH	N/A	N/A	6B73, 6VI4	N/A
μ-OR	N/A	N/A	N/A	6DDF, 6DDE	N/A
NOPR	N/A	N/A	N/A	N/A	5DHH, 5DHG, 4EA3
OX$_1$R	6TP3, 6TOT, 6TP4, 6TP6, 6TQ4, 6TQ6, 6TQ7, 6TQ9, 6TOD, 6TO7, 6TOS	N/A	N/A	N/A	4ZJC, 4ZJ8, 6V9S
OX$_2$R	6TPN, 6TPG	7L1V, 7L1U	N/A	N/A	4S0V, 5WQC, 5WS3
PAR1	3VW7	N/A	N/A	N/A	N/A
PAR2	5NDD, 5NDZ, 5NJ6	N/A	N/A	N/A	N/A
NK$_1$R	6HLL, 6J21, 6J20, 6HLO, 6HLP	N/A	N/A	N/A	6E59
V$_{1A}$R	1YTV	N/A	N/A	N/A	N/A
V$_2$R	N/A	N/A	N/A	N/A	6U1N, 6NI2
OTR	6TPK	N/A	N/A	N/A	N/A
ACKR$_1$	4NUU, 4NUV	N/A	N/A	N/A	N/A

(continued)

Table 1.2 (continued)

Receptor	R (inactive state)	R' (inactive agonist-bound state)	R'' (active state)	R* (active state, with G protein subunit)	R*G (G protein signaling state)
FFA$_1$R	N/A	4PHU, 5TZY, 5TZR	5KW2	N/A	N/A
CysLT$_1$R	6RZ5, 6RZ4	N/A	N/A	N/A	N/A
CysLT$_2$R	6RZ6, 6RZ7, 6RZ8, 6RZ9	N/A	N/A	N/A	N/A
LPA$_1$R	4Z34, 4Z35, 4Z36	N/A	N/A	N/A	N/A
S$_1$P$_1$R	3V2W, 3V2Y	N/A	N/A	N/A	N/A
CB$_1$R	5ZTY, 6KPC, 6KPF	N/A	N/A	6PT0	N/A
PAFR	N/A	N/A	N/A	N/A	5ZKQ, 5ZKP
DP$_1$R	6D26, 6D27	N/A	N/A	N/A	N/A
EP$_2$R	N/A	N/A	N/A	7CX2, 7CX3, 7CX4	N/A
EP$_3$R	N/A	N/A	6AK3, 6M9T	N/A	N/A
EP$_4$R	N/A	N/A	N/A	7D7M	5YWY, 5YHL
TPR	N/A	N/A	N/A	N/A	6IIU, 6IIV
A$_1$R	5UEN, 5N2S	N/A	N/A	N/A	6D9H
P$_2$Y$_1$R	4XNV, 4XNW	N/A	N/A	N/A	N/A
P$_2$Y$_{12}$R	N/A	4PXZ, 4PY0, 4NTJ	N/A	N/A	N/A
GPBAR	N/A	N/A	7BW0	7CFM, 7CFN	N/A
GPR52	N/A	N/A	N/A	6LI0, 6LI1, 6LI2, 6LI3	N/A
MT$_1$R	N/A	6PS8	6ME2, 6ME3, 6ME4, 6ME5	N/A	N/A
MT$_2$R	N/A	N/A	6ME6, 6ME7, 6ME8, 6ME9	N/A	N/A
Ghs-R	N/A	N/A	N/A	6KO5	N/A

(continued)

Table 1.2 (continued)

Receptor	R (inactive state)	R' (inactive agonist-bound state)	R'' (active state)	R* (active state, with G protein subunit)	R*G (G protein signaling state)
CCR$_2$	N/A	N/A	N/A	N/A	6GPX, 6GPS, 5T1A
CCR$_5$	N/A	N/A	N/A	6MEO, 6MET	4MBS, 5UIW, 6AKX, 6AKY
CCR$_6$	N/A	N/A	N/A	6WWZ	N/A
CCR$_7$	N/A	N/A	N/A	N/A	6QZH
CCR$_9$	5LWE	N/A	N/A	N/A	N/A
CXCR$_2$	N/A	N/A	N/A	6LFL, 6KVF, 6KVA	N/A
CXCR$_4$	N/A	N/A	N/A	N/A	3OE9, 3ODU, 3OE8, 3OE0, 3OE6, 4RWS
FZD$_4$	6BD4	N/A	N/A	N/A	5CL1, 6NE1
FZD$_5$	N/A	N/A	N/A	6WW2	6O39, 5URY, 5URZ
SMO	4JKV, 4N4W, 5V57, 5V56, 5L7I, 5L7D	N/A	N/A	6XBK, 6XBJ, 6XBM, 6XBL	4QIM, 4O9R, 4QIN, 6OT0
CTR	N/A	N/A	N/A	6NIY	6PFO, 6PGQ, 5UZ7
CTLR	6ZHO, 6ZIS, 6V2E	N/A	N/A	4RWF, 4RWG, 5V6Y, 6UMG	7KNT, 7KNU
CRF$_1$R	4K5Y, 3EHS	N/A	N/A	6P9X, 3EHT, 3EHU	4Z9G, 6PB0
CRF$_2$R	N/A	N/A	N/A	N/A	6PB1
GHRHR	N/A	N/A	N/A	7CZ5	N/A
GLP$_1$R	N/A	7LCK	5VAI, 5NX2, 7C2E, 6XOX, 6X1A, 6X18, 6X19	6ORV, 7LCJ	6B3J, 6KK1, 5VEW, 5VEX, 6KJV, 6KK7, 6VCB

(continued)

Table 1.2 (continued)

Receptor	R (inactive state)	R′ (inactive agonist-bound state)	R″ (active state)	R* (active state, with G protein subunit)	R*G (G protein signaling state)
GLP$_2$R	N/A	N/A	N/A	7D68	N/A
GCGR	4L6R	N/A	5NX2	6WHC	5EE7, 5XEZ, 5XF1, 5YQZ, 6WPW, 5VEW, 5VEX, 6LML, 6LMK, 6B3J
SR	N/A	N/A	N/A	7D3S, 6WZG, 6WI9	N/A
PTH$_1$R	N/A	N/A	6FJ3	N/A	6NBF, 6NBI, 6NBH
PAC$_1$R	N/A	N/A	N/A	6P9Y	6LPB
VPAC$_1$R	N/A	N/A	N/A	N/A	6VN7

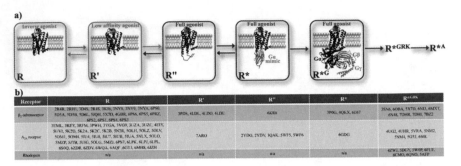

Fig. 1.1 Intermediates in the G protein-coupled receptor (GPCR) activation mechanism. **a** A schematic diagram. **b** The crystal structures of the three most well-studied human GPCRs (taking β_2-adrenoceptor as an example) were obtained from Protein Data Bank. R: inactive state (2RH1); R′: inactive low-affinity agonist-bound state (5JQH); R″: active state (6KR8); R*: activated state with Gα or surrogate mimic g (3P0G); R*^{G}: G protein signaling conformation (3SN6); R*^{GRK}: GPCR binding to G protein receptor kinase; R*^{A}: GPCR binding to β-arrestin

will be achieved in the presence of heterotrimeric G protein. The drastically distinct distribution between these ligand-free receptor states is promising to reflect their diverse level of basal activities (Katritch et al. 2013). Through the comprehensive characterization of conformational changes at the intracellular side, GPCRs have been approved to shift the equilibrium toward the activated states after their binding to specific agonists (Frei et al. 2020). Oppositely, the binding of inverse agonists induces the equilibrium toward inactive states, which results in a loss of receptor activity (Nygaard et al. 2013). Neutral antagonists have little effect on the basal equilibrium. Although a grand contribution to profile GPCR basal activity, these conformational states are far away from the requirements of the research on the GPCR signaling pathway.

As an impressive strategy for realizing protein dynamic properties, classical molecular dynamics simulations are limited to the timescales that are orders of magnitude shorter than the conformational changes relevant for most biological functions, in particular, when it comes to increasing crystal structures of GPCRs. In a case to discrete all the conformations between β_2-AR 'on' and 'off' states for drug discovery, such an issue has been addressed by collaboration between Stanford University and Google Company. Taking inspiration from Markov State Models, they have developed a new method for mimicking the changes of GPCR three-dimensional structures at the atom level. Cloud-based simulations on Google Exacycle using this method have provided an atomistic description of the activation of a GPCR and reveal multiple activation pathways (Kohlhoff et al. 2014).

Crystallography typically captures the lowest energy states within an ensemble of conformations. Other methods are therefore required to characterize transiently populated conformations as well as the transitions between different conformations. Using NMR spectroscopy of $^{13}CH_3$-ε-methionines, the work by Kofuku et al.

observed significant β_2-AR conformational heterogeneity of agonist- and inverse-bound states in the transmembrane region (Kofuku et al. 2012). Further study using ^{19}F-fluorine NMR and double electron-electron resonance spectroscopy has proved that unliganded and inverse agonist-bound β_2-AR exists predominantly in two inactive conformations that exchange dynamically within hundreds of microseconds (Nygaard et al. 2013). The binding of agonists partially shifts the equilibrium toward a conformation having the capacity to engage G proteins whereby results in increased conformational heterogeneity and the coexistence of inactive, intermediate, and active states. Subsequent coupling to G protein or an intracellular G protein mimetic induces a complete transition to the active conformation (Manglik et al. 2015).

1.3.2 G Protein-Coupled Receptor Signaling Pathway

The widely accepted paradigm for GPCR activation involves the occupancy of the receptor by specific ligands, the initiation of transient activation, and the receptor recycling for another round of stimulation by inactivation. This cyclical model has been predicated both on conceptual grounds and technical limitations. Such issues have been addressed by Fluorescence Resonance Energy transfer (FRET)-based sensors (Irannejad et al. 2013) and enhanced Bioluminescence Resonance Energy Transfer (BRET) procedures. Successful application of the two assays permits real-time observations of signaling and trafficking events in live cells and measurement of GPCR activation and trafficking (Namkung et al. 2016; Stoddart et al. 2015).

Activation of GPCR by their traditional agonists promotes conformational changes followed by activating at least three families of proteins in an agonist-dependent manner. The three families include the heterotrimeric G proteins, GPCR kinases/GRKs, and arrestins (Campbell and Smrcka 2018). Regarding the sequence similarity of the $G\alpha$ subunit, the heterotrimeric G proteins have been generally divided into four main classes: $G\alpha_s$, $G\alpha_{i/o}$, $G\alpha_{q/11}$, and $G\alpha_{12/13}$ (Fig. 1.2a). In an activated conformation, GPCRs function as efficient guanine nucleotide exchange factors (GEFs) for bound G proteins and promote the activation of $G\alpha$ and their dissociation from $G\beta\gamma$ subunits. The activated G protein subunits in turn trigger a signal transduction cascade through diverse effectors. Both the $G\alpha_s$ and $G\alpha_{i/o}$ pathways are mediated by the adenylate cyclase (AC) that generates cyclic-adenosine monophosphate (cAMP). Despite distinct gene products with subtle differences in tissue distribution or function, all the ten reported ACs are directly stimulated by G-proteins of the $G\alpha s$ class, whereby exhibit the capacity to catalyze the conversion of cytosolic adenosine triphosphate (ATP) to cAMP. In contrast, interaction with $G\alpha$ subunits of the $G\alpha_{i/o}$ type inhibits AC from generating cAMP. The cAMP is regarded as a second messenger owing to its significant role in regulating the activity of diverse ion channels and the effector-like members of the ser/thr-specific protein kinase A (PKA) family. Phospholipase C-β (PLCβ) serves as the effector of the $G\alpha_{q/11}$ pathway by catalyzing the cleavage of membrane-bound phosphatidylinositol 4,5-bisphosphate (PIP2) into the second messengers inositol trisphosphate

(IP3) and diacylglycerol (DAG) (Kankanamge et al. 2021). The interaction between IP3 and IP3 receptors elicits Ca^{2+} release from the endoplasmic reticulum, while the diffusion of DAG activates a second ser/thr kinase (protein kinase C, PKC). The two pathways have proved to converge on each other to signal through the same secondary effector because the growth of intracellular Ca^{2+} often results in the activation of most PKC isoforms. The enhanced intracellular Ca^{2+} is reported to bind and allosterically activate calmodulins which further target and allosterically activate enzymes such as Ca^{2+}/calmodulin-dependent kinases (CAMKs). The $G\alpha_{12/13}$ pathway has three effectors identified as RhoGEFs (p115-RhoGEF, PDZ-RhoGEF, and LARG). The occupancy of the three effectors on $G\alpha_{12/13}$ allosterically activates Rho family of GTPases. The binding of Rho to GTP continuously activates downstream proteins that are responsible for cytoskeleton regulation such as Rho-kinase (ROCK). In most cases, the GPCRs that interact with $G\alpha_{12/13}$ are able to target $G\alpha_{q/11}$. The primary effectors of $G\beta\gamma$ are distinct ion channels: G protein-regulated inwardly rectifying K^+ channels (GIRKs), P/Q- and N-type voltage-gated Ca^{2+} channels, several isoforms of AC and PLC, as well as some phosphoinositide-3-kinase (PI3K) isoforms. Although great progress in the G protein-dependent signaling pathway, it is clear that considerably more remains to be explored about the molecular dynamics of these fundamental activation mechanisms and their operations in living cells (Gregorio et al. 2017; Sungkaworn et al. 2017).

Recent studies have evidenced that several GPCRs signal without the presence of G proteins, and heterotrimeric G proteins sometimes function independently with GPCRs. Many proteins mentioned in G protein-dependent signaling plays important role in the G protein-independent pathway, for instance, GRKs and β-arrestins (Fig. 1.2b). Although both GRKs and β-arrestins are initially identified as inhibitors of GPCR signaling, their roles in GPCR signal transduction are multifactorial and continuously evolving. Phosphorylation of activated GPCRs by GRKs, originally documented as the first step in GPCR desensitization is now described as a critical regulatory mechanism. Such mechanism triggers diverse downstream effects by engaging different conformations of β-arrestins and promoting their various functions (Reiter et al. 2011; Lee et al. 2016; Yang et al. 2015; Nuber et al. 2016). Studies on many GPCRs (dopamine D1 receptor (Kaya et al. 2020), follicle-stimulating hormone receptor (Kara et al. 2006), β_2-AR (Krasel et al. 2008), thyrotropin-releasing hormone receptor (Jones and Hinkle 2008), somatostatin 2A receptor (Pöll et al. 2011), neuropeptide FF2 receptor (Bray et al. 2012) and free fatty acid receptor 4 (Senatorov et al. 2020)) have confirmed that the elimination of specific phosphorylation sites in a GPCR has distinct downstream effects on β-arrestin binding and/or signaling. These reports, together with the studies on M_3R (Poulin et al. 2010) and μ-opioid receptors (Mafi et al. 2020) support the idea that sensitization of distinct phosphorylation motifs by GRKs yields various outcomes through subsequent protein interactions, especially via β-arrestin recruitment.

Recent extensive reports have changed our appreciation of the functional roles of β-arrestins. β-arrestins have proved to control the trafficking itinerary of internalized GPCRs by functioning as critical adaptors for agonist-induced endocytosis and agonist-induced ubiquitination of GPCRs (Ferguson et al. 1996; Goodman et al.

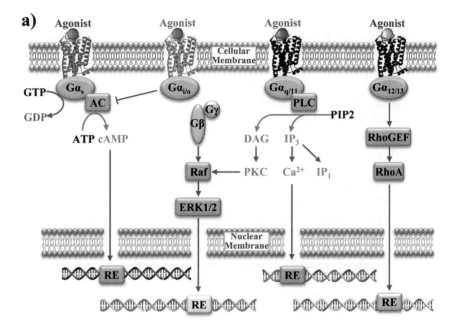

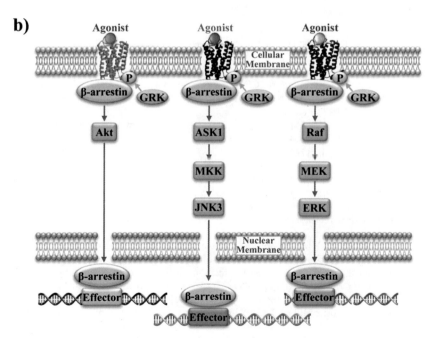

Fig. 1.2 GPCR signaling pathway. **a** G protein signaling network; **b** arrestin signaling network

1996; Han et al. 2012a, b; Jean-Charles et al. 2016; Shenoy et al. 2001, 2008). Second, β-arrestins are capable of forming signaling scaffolds for mitogen-activated protein kinases (MAPKs) such as the extracellular signaling kinases 1 and 2 (ERK), and c-Jun N-terminal kinase 3 (JNK3) on endosomes with internalized GPCRs (Luttrell and Miller 2013). Finally, β-arrestins have the capacity to promote some of these pathways even under the condition that the G protein activity is disabled (Han et al. 2012a, b; Whalen et al. 2010). Taken together, GPCR signaling is bimodal whereby the first mode is dependent on G proteins and the second mode is dependent on GRK phosphorylation of the GPCR and β-arrestin-dependent signaling (Lefkowitz and Shenoy 2005; Violin and Lefkowitz 2007; Luttrell 2014; Hodavance et al. 2016).

Although tremendous strides have been made in realizing GPCR signaling, key issues in physiological situations and pharmacological interventions remain to be addressed. These include the mechanism of signaling from internal compartments distinguishing examples of prolonged signaling at the cell surface; the spatiotemporal logic of producing rapidly diffusing messengers (cAMP) at two locations; the co-trafficking of downstream G proteins, cyclases, arrestins, and the structural and conformational dynamics whereby the protein trafficking hand-offs take place.

References

Adikesavan NV et al (2005) AC-terminal segment of the V1R vasopressin receptor is unstructured in the crystal structure of its chimera with the maltose-binding protein. Acta Crystallogr Sect F Struct Biol Cryst Commun 61(Pt 4):341–345. https://doi.org/10.1107/S1744309105007293

Apel AK et al (2019) Crystal structure of CC chemokine receptor 2A in complex with an orthosteric antagonist provides insights for the design of selective antagonists. Structure 27(3):427-438.e5. https://doi.org/10.1016/j.str.2018.10.027

Asada H et al (2020) The crystal structure of angiotensin II type 2 receptor with endogenous peptide hormone. Structure 28(4):418-425.e4. https://doi.org/10.1016/j.str.2019.12.003

Batchelor JD et al (2014) Red blood cell invasion by Plasmodium vivax: structural basis for DBP engagement of DARC. PLoS Pathog 10(1):e1003869. https://doi.org/10.1371/journal.ppat

Bockaert J, Pin JP (1999) Molecular tinkering of G protein-coupled receptors: an evolutionary success. EMBO J 18(7):1723–1729. https://doi.org/10.1093/emboj/18.7.1723

Bray L et al (2012) Identification and functional characterization of the phosphorylation sites of the neuropeptide FF2 receptor. J Biol Chem 289(49):33754–33766. https://doi.org/10.1074/jbc.M114.612614

Calebiro D et al (2021) G protein-coupled receptor-G protein interactions: a single-molecule perspective. Physiol Rev 101(3):857–906. https://doi.org/10.1152/physrev.00021

Campbell A, Smrcka A (2018) Targeting G protein-coupled receptor signalling by blocking G proteins. Nat Rev Drug Discov 17(11):789–803. https://doi.org/10.1038/nrd.2018.135

Cao C et al (2018) Structural basis for signal recognition and transduction by platelet-activating-factor receptor. Nat Struct Mol Biol 25(6):488–495. https://doi.org/10.1038/s41594-018-0068-y

Che T et al (2017) Structure of the nanobody-stabilized active state of the kappa opioid receptor. Cell 172(1–2):55-67.e15. https://doi.org/10.1016/j.cell.2017.12.011

Chen T et al (2020) Structural basis of ligand binding modes at the human formyl peptide receptor 2. Nat Commun 11:1208. https://doi.org/10.1038/s41467-020-15009-1

Chen L, Jin L, Zhou N (2012) An update of novel screening methods for GPCR in drug discovery. Expert Opin Drug Discov 7(9):791–806. https://doi.org/10.1517/17460441.2012.699036

Cheng R et al (2017) Structural insight into allosteric modulation of protease-activated receptor 2. Nature 545(7652):112–115. https://doi.org/10.1038/nature22309

Cherezov V et al (2007) High-resolution crystal structure of an engineered human beta2-adrenergic G protein-coupled receptor. Science 318(5854):1258–1265. https://doi.org/10.1126/science.1150577

Chien EY et al (2010) Structure of the human dopamine D3 receptor in complex with a D2/D3 selective antagonist. Science 330(6007):1091–1095. https://doi.org/10.1126/science.1197410

Chrencik JE et al (2015) Crystal structure of antagonist bound human lysophosphatidic acid receptor 1. Cell 161(7):1633–1643. https://doi.org/10.1016/j.cell.2015.06.002

Christopher JA et al (2019) Structure-based optimization strategies for G protein-coupled receptor (GPCR) allosteric modulators: a case study from analyses of new metabotropic glutamate receptor 5 (mGlu$_5$) X-ray structures. J Med Chem 62(1):207–222. https://doi.org/10.1021/acs.jmedchem.7b01722

Colley CS et al (2018) Structure and characterization of a high affinity C5a monoclonal antibody that blocks binding to C5$_a$R$_1$ and C5$_a$R$_2$ receptors. Mabs 10(1):104–117. https://doi.org/10.1080/19420862.2017.1384892

Cornwell AC, Feigin ME (2020) Unintended effects of GPCR-targeted drugs on the cancer phenotype. Trends Pharmacol Sci 41(12):1006–1022. https://doi.org/10.1016/j.tips.2020.10.001

Dong M et al (2020) Structure and dynamics of the active Gs-coupled human secretin receptor. Nat Commun 11(1):4137. https://doi.org/10.1038/s41467-020-17791-4

Douthwaite JA et al (2015) Affinity maturation of a novel antagonistic human monoclonal antibody with a long VH CDR$_3$ targeting the Class A GPCR formyl-peptide receptor 1. Mabs 7(1):152–166. https://doi.org/10.4161/19420862.2014.985158

Emtage AL et al (2017) GPCRs through the keyhole: the role of protein flexibility in ligand binding to β-adrenoceptors. J Biomol Struct Dyn 35(12):2604–2619. https://doi.org/10.1080/07391102.2016.1226197

Fan H et al (2019) Structural basis for ligand recognition of the human thromboxane A2 receptor. Nat Chem Biol 15(1):27–33. https://doi.org/10.1038/s41589-018-0170-9

Fan L et al (2020) Haloperidol bound D2 dopamine receptor structure inspired the discovery of subtype selective ligands. Nat Commun 11:1074. https://doi.org/10.1038/s41467-020-14884-y

Fenalti G et al (2014) Molecular control of δ-opioid receptor signalling. Nature 506(7487):191–196. https://doi.org/10.1038/nature12944

Feng G et al (2020) G protein-coupled receptor-in-paper, a versatile chromatographic platform to study receptor-drug interaction. J Chromatogr A 1637:461835. https://doi.org/10.1016/j.chroma.2020.461835

Ferguson SS et al (1996) Role of beta-arrestin in mediating agonist-promoted G protein-coupled receptor internalization. Science 271(5247):363–366. https://doi.org/10.1126/science.271.5247.363

Fredriksson R et al (2003). The G-protein-coupled receptors in the human genome form five main families. Phylogenetic analysis, paralogon groups, and fingerprints. Mol Pharmacol 63(6):1256–1272. https://doi.org/10.1124/mol.63.6.1256

Frei JN et al (2020) Conformational plasticity of ligand-bound and ternary GPCR complexes studied by 19F NMR of the β1-adrenergic receptor. Nat Commun 11(1):669. https://doi.org/10.1038/s41467-020-14526-3

Früh V, IJzerman AP, Siegal G (2011) How to catch a membrane protein in action: a review of functional membrane protein immobilization strategies and their applications. Chem Rev 111(2):640–656. https://doi.org/10.1021/cr900088s

Glukhova A et al (2017) Structure of the adenosine A1 receptor reveals the basis for subtype selectivity. Cell 168(5):867-877.e13. https://doi.org/10.1016/j.cell.2017.01.042

Goodman O et al (1996) β-arrestin acts as a clathrin adaptor in endocytosis of the β2-adrenergic receptor. Nature 383(6599):447–450. https://doi.org/10.1038/383447a0

Graul RC, Sadée W (2001) Evolutionary relationships among G protein-coupled receptors using a clustered database approach. AAPS PharmSci 3(2):E12. https://doi.org/10.1208/ps030212

Gregorio G et al (2017) Single-molecule analysis of ligand efficacy in β2AR–G-protein activation. Nature 547(7661):68–77. https://doi.org/10.1038/nature22354

Gusach A et al (2019) Structural basis of ligand selectivity and disease mutations in cysteinyl leukotriene receptors. Nat Commun 10(1):5573. https://doi.org/10.1038/s41467-019-13348-2

Haga K et al (2012) Structure of the human M2 muscarinic acetylcholine receptor bound to an antagonist. Nature 482(7386):547–551. https://doi.org/10.1038/nature10753

Han SO et al (2012a) MARCH2 promotes endocytosis and lysosomal sorting of carvedilol-bound β(2)-adrenergic receptors. J Cell Biol 199(5):817–830. https://doi.org/10.1083/jcb.201208192

Han SO, Kommaddi RP, Shenoy SK (2012b) Distinct roles for β-arrestin2 and arrestin-domain-containing proteins in β2 adrenergic receptor trafficking. EMBO Rep 14(2):164–171. https://doi.org/10.1038/embor.2012.187

Hanson MA et al (2012) Crystal structure of a lipid G protein-coupled receptor. Science 335(6070):851–855. https://doi.org/10.1126/science.1215904

Hauser A et al (2017) Trends in GPCR drug discovery: new agents, targets and indications. Nat Rev Drug Discov 16:829–842. https://doi.org/10.1038/nrd.2017.178

Hodavance SY et al (2016) G protein-coupled receptor biased agonism. J Cardiovasc Pharmacol 67(3):193–202. https://doi.org/10.1097/FJC.0000000000000356

Huang W et al (2020) Structure of the neurotensin receptor 1 in complex with β-arrestin 1. Nature 579:303–308. https://doi.org/10.1038/s41586-020-1953-1

Irannejad R et al (2013) Conformational biosensors reveal GPCR signalling from endosomes. Nature 495(7442):534–538. https://doi.org/10.1038/nature12000

Jaakola VP et al (2008) The 2.6 angstrom crystal structure of a human A_2A adenosine receptor bound to an antagonist. Science 322(5905):1211–1217. https://doi.org/10.1126/science.1164772

Jaeger K et al (2019) Standfuss J. Structural basis for allosteric ligand recognition in the human CC chemokine receptor 7. Cell 178(5):1222–1230.e10. https://doi.org/10.1016/j.cell.2019.07.028

Joost P, Methner A (2002) Phylogenetic analysis of 277 human G-protein-coupled receptors as a tool for the prediction of orphan receptor ligands. Genome Biol 3(11):research0063. https://doi.org/10.1186/gb-2002-3-11-research0063

Jean-Charles PY, Snyder JC, Shenoy SK (2016) Chapter one—Ubiquitination and deubiquitination of G protein-coupled receptors. Prog Mol Biol Transl Sci 141:1–55. https://doi.org/10.1016/bs.pmbts.2016.05.001

Johansson LC et al (2019) XFEL structures of the human MT_2 melatonin receptor reveal the basis of subtype selectivity. Nature 569(7755):289–292. https://doi.org/10.1038/s41586-019-1144-0

Jones BW, Hinkle PM (2008) Arrestin binds to different phosphorylated regions of the thyrotropin-releasing hormone receptor with distinct functional consequences. Mol Pharmacol 74(1):195–202. https://doi.org/10.1124/mol.108.045948

Josefsson LG (1999) Evidence for kinship between diverse G-protein coupled receptors. Gene 239(2):333–340. https://doi.org/10.1016/s0378-1119(99)00392-3

Kang Y et al (2015) Crystal structure of rhodopsin bound to arrestin by femtosecond X-ray laser. Nature 523(7562):561–567. https://doi.org/10.1038/nature14656

Kankanamge D et al (2021) Dissociation of the G protein βγ from the Gq-PLCβ complex partially attenuates PIP_2 hydrolysis. J Biol Chem 296:100702. https://doi.org/10.1016/j.jbc.2021.100702

Kara E et al (2006) A phosphorylation cluster of five serine and threonine residues in the C-terminus of the follicle-stimulating hormone receptor is important for desensitization but not for β-arrestin-mediated ERK activation. Mol Endocrinol 20(11):3014–3026. https://doi.org/10.1210/me.2006-0098

Katritch V, Cherezov V, Stevens RC (2013) Structure-function of the G protein-coupled receptor superfamily. Annu Rev Pharmacol Toxicol 53:531–556. https://doi.org/10.1146/annurev-pharmtox-032112-135923

Kaya AI et al (2020) Phosphorylation barcode-dependent signal bias of the dopamine D1 receptor. Proc Natl Acad Sci 117(25):14139–14149. https://doi.org/10.1073/pnas.1918736117

Kimura KT et al (2019) Structures of the 5-HT$_{2A}$ receptor in complex with the antipsychotics risperidone and zotepine. Nat Struct Mol Biol 26(2):121–128. https://doi.org/10.1038/s41594-018-0180-z

Kobilka BK (2007) G protein coupled receptor structure and activation. Biochim Biophys Acta 1768(4):794–807. https://doi.org/10.1016/j.bbamem.2006.10.021

Koehl A et al (2018) Structure of the μ-opioid receptor–G$_i$ protein complex. Nature 558(7711):547–552. https://doi.org/10.1038/s41586-018-0219-7

Kofuku Y et al (2012) Efficacy of the β2-adrenergic receptor is determined by conformational equilibrium in the transmembrane region. Nat Commun 3:1045. https://doi.org/10.1038/ncomms2046

Kohlhoff K et al (2014) Cloud-based simulations on Google Exacycle reveal ligand modulation of GPCR activation pathways. Nature Chem 6(1):15–21. https://doi.org/10.1038/nchem.1821

Kolakowski LF Jr (1994) GCRDb: a G-protein-coupled receptor database. Recept Channels 2(1):1–7

Krasel C et al (2008) Dual role of the β2-adrenergic receptor C terminus for the binding of beta-arrestin and receptor internalization. J Biol Chem 283(46):31840–33188. https://doi.org/10.1074/jbc.M806086200

Krishnan A et al (2014) The GPCR repertoire in the demosponge Amphimedon queenslandica: insights into the GPCR system at the early divergence of animals. BMC Evol Biol 14:270. https://doi.org/10.1186/s12862-014-0270-4

Lebon G et al (2011) Agonist-bound adenosine A2A receptor structures reveal common features of GPCR activation. Nature 474(7352):521–525. https://doi.org/10.1038/nature10136

Lee MH et al (2016) The conformational signature of β-arrestin2 predicts its trafficking and signalling functions. Nature 531(7596):665–668. https://doi.org/10.1038/nature17154

Lefkowitz RJ, Shenoy SK (2005) Transduction of receptor signals by beta-arrestins. Science 308(5721):512–517. https://doi.org/10.1126/science.1109237

Li X et al (2019) Crystal structure of the human cannabinoid receptor CB$_2$. Cell 176(3):459–467 e13. https://doi.org/10.1016/j.cell.2018.12.011

Lin X et al (2020) Structural basis of ligand recognition and self-activation of orphan GPR52. Nature 579(7797):152–157. https://doi.org/10.1038/s41586-020-2019-0

Liu K et al (2020) Structural basis of CXC chemokine receptor 2 activation and signalling. Nature 585(7823):135–140. https://doi.org/10.1038/s41586-020-2492-5

Luginina A et al (2019) Structure-based mechanism of cysteinyl leukotriene receptor inhibition by antiasthmatic drugs. Sci Adv 5(10):eaax2518. https://doi.org/10.1126/sciadv.aax2518

Luttrell LM (2014) Minireview: more than just a hammer: ligand "bias" and pharmaceutical discovery. Mol Endocrinol 28(3):281–294. https://doi.org/10.1210/me.2013-1314

Luttrell LM, Miller WE (2013) Arrestins as regulators of kinases and phosphatases. Prog Mol Biol Transl Sci 118:115–147. https://doi.org/10.1016/B978-0-12-394440-5.00005-X

Ma Y et al (2017a) Structural basis for apelin control of the human apelin receptor. Structure 25(6):858–866.e4. https://doi.org/10.1016/j.str.2017.04.008

Ma P et al (2017b) The cubicon method for concentrating membrane proteins in the cubic mesophase. Nat Protoc 12(9):1745–1762. https://doi.org/10.1038/nprot.2017.057

Maeda S et al (2020) Structure and selectivity engineering of the M1 muscarinic receptor toxin complex. Science 369(6500):161–167. https://doi.org/10.1126/science.aax2517

Mafi A, Kim SK, Goddard WA 3rd (2020) Mechanism of β-arrestin recruitment by the μ-opioid G protein-coupled receptor. Proc Natl Acad Sci 117(28):16346–16355. https://doi.org/10.1073/pnas.1918264117

Manglik A et al (2015) Structural insights into the dynamic process of β2-adrenergic receptor signaling. Cell 161(5):1101–1111. https://doi.org/10.1016/j.cell.2015.04.043

Miller RL et al (2015) The importance of ligand-receptor conformational pairs in stabilization: spotlight on the N/OFQ G protein-coupled receptor. Structure 23(12):2291–2299. https://doi.org/10.1016/j.str.2015.07.024

Mombaerts P (2004) Genes and ligands for odorant, vomeronasal and taste receptors. Nat Rev Neurosci 5(4):263–278. https://doi.org/10.1038/nrn1365

Morimoto K et al (2019) Crystal structure of the endogenous agonist-bound prostanoid receptor EP3. Nat Chem Biol 15(1):8–10. https://doi.org/10.1038/s41589-018-0171-8

Nagiri C et al (2019) Crystal structure of human endothelin ETB receptor in complex with peptide inverse agonist IRL2500. Commun Biol 2:236. https://doi.org/10.1038/s42003-019-0482-7

Namkung Y et al (2016) Monitoring G protein-coupled receptor and β-arrestin trafficking in live cells using enhanced bystander BRET. Nat Commun 7:12178. https://doi.org/10.1038/ncomms 12178

Nuber S et al (2016) β-arrestin biosensors reveal a rapid, receptor-dependent activation/deactivation cycle. Nature 531(7596):661–664. https://doi.org/10.1038/nature17198

Nygaard R et al (2013) The dynamic process of β(2)-adrenergic receptor activation. Cell 152(3):532– 542. https://doi.org/10.1016/j.cell.2013.01.008

Oswald C et al (2016) Intracellular allosteric antagonism of the CCR9 receptor. Nature 540(7633):462–465. https://doi.org/10.1038/nature20606

Overington J, Al-Lazikani B, Hopkins A (2006) How many drug targets are there? Nat Rev Drug Discov 5(12):993–996. https://doi.org/10.1038/nrd2199

Peng Y et al (2018) 5-HT_{2C} receptor structures reveal the structural basis of GPCR polypharmacology. Cell 172(4):719–730. e14. https://doi.org/10.1016/j.cell.2018.01.001

Pfleger J, Gresham K, Koch WJ (2019) G protein-coupled receptor kinases as therapeutic targets in the heart. Nat Rev Cardiol 16(12):612–662. https://doi.org/10.1038/s41569-019-0220-3

Pöll F, Doll C, Schulz S (2011) Rapid dephosphorylation of G protein-coupled receptors by protein phosphatase 1β is required for termination of β-arrestin-dependent signaling. J Biol Chem 286(38):32931–32936. https://doi.org/10.1074/jbc.M111.224899

Poulin B et al (2010) The M3-muscarinic receptor regulates learning and memory in a receptor phosphorylation/arrestin-dependent manner. Proc Natl Acad Sci 107(20):9440–9445. https://doi.org/10.1073/pnas.0914801107

Qin L et al (2015) Structural biology. Crystal structure of the chemokine receptor CXCR4 in complex with a viral chemokine. Science 347(6226):1117–1122. https://doi.org/10.1126/science.1261064

Qu C et al (2021) Ligand recognition, unconventional activation, and G protein coupling of the prostaglandin E2 receptor EP2 subtype. Sci Adv 7(14):eabf1268. https://doi.org/10.1126/sciadv.abf1268

Reiter E et al (2011) Molecular mechanism of β-arrestin-biased agonism at seven-transmembrane receptors. Annu Rev Pharmacol Toxicol 52:179–197. https://doi.org/10.1146/annurev.pharmtox.010909.105800

Rasmussen S et al (2007) Crystal structure of the human β2 adrenergic G-protein-coupled receptor. Nature 450(7168):383–387. https://doi.org/10.1038/nature06325

Robertson N et al (2018) Structure of the complement C5a receptor bound to the extra-helical antagonist NDT9513727. Nature 553(7686):111–114. https://doi.org/10.1038/nature25025

Rosenbaum MI et al (2020) Targeting receptor complexes: a new dimension in drug discovery. Nat Rev Drug Discov 19(12):884–901. https://doi.org/10.1038/s41573-020-0086-4

Rovati GE, Capra V, Neubig RR (2007) The highly conserved DRY motif of class A G protein-coupled receptors: beyond the ground state. Mol Pharmacol 71(4):959–964. https://doi.org/10.1124/mol.106.029470

Russ AP, Lampel S (2005) The druggable genome: an update. Drug Discov Today 10(23–24):1607–1610. https://doi.org/10.1016/S1359-6446(05)03666-4

Schöppe J et al (2019) Crystal structures of the human neurokinin 1 receptor in complex with clinically used antagonists. Nat Commun 10(1):17. https://doi.org/10.1038/s41467-018-07939-8

Schwartz TW et al (2006) Molecular mechanism of 7TM receptor activation–a global toggle switch model. Annu Rev Pharmacol Toxicol 46:481–519. https://doi.org/10.1146/annurev.pharmtox.46.120604.141218

Senatorov IS et al (2020) Carboxy-terminal phosphoregulation of the long splice isoform of free-fatty acid receptor-4 mediates β-arrestin recruitment and signaling to $ERK_{1/2}$. Mol Pharmacol 97(5):304–313. https://doi.org/10.1124/mol.119.117697

Shenoy SK et al (2001) Regulation of receptor fate by ubiquitination of activated beta 2-adrenergic receptor and beta-arrestin. Science 294(5545):1307–1313. https://doi.org/10.1126/science.106 3866

Shenoy SK et al (2008) Nedd4 mediates agonist-dependent ubiquitination, lysosomal targeting, and degradation of the beta2-adrenergic receptor. J Biol Chem 283(32):22166–22176. https://doi.org/ 10.1074/jbc.M709668200

Shiimura Y et al (2020) Structure of an antagonist-bound ghrelin receptor reveals possible ghrelin recognition mode. Nat Commun 11(1):4160. https://doi.org/10.1038/s41467-020-17554-1

Shimamura T et al (2011) Structure of the human histamine H_1 receptor complex with doxepin. Nature 475:65–70. https://doi.org/10.1038/nature10236

Smith MT et al (2013) Enhanced protein stability through minimally invasive, direct, covalent, and site-specific immobilization. Biotechnol Prog 29(1):247–254. https://doi.org/10.1002/btpr.1671

Stauch B et al (2019) Structural basis of ligand recognition at the human MT_1 melatonin receptor. Nature 569(7755):284–288. https://doi.org/10.1038/s41586-019-1141-3

Staus DP et al (2020) Structure of the M_2 muscarinic receptor–β-arrestin complex in a lipid nanodisc. Nature 579(7798):297–302. https://doi.org/10.1038/s41586-020-1954-0

Stoddart L et al (2015) Application of BRET to monitor ligand binding to GPCRs. Nat Methods 12(7):661–663. https://doi.org/10.1038/nmeth.3398

Sungkaworn T et al (2017) Single-molecule imaging reveals receptor–G protein interactions at cell surface hot spots. Nature 550(7677):543–547. https://doi.org/10.1038/nature24264

Suno R et al (2018) Structural insights into the subtype-selective antagonist binding to the M2 muscarinic receptor. Nat Chem Biol 14(12):1150–1158. https://doi.org/10.1038/s41589-018-0152-y

Tan Q et al (2013) Structure of the CCR_5 chemokine receptor-HIV entry inhibitor maraviroc complex. Science 341(6152):1387–1390. https://doi.org/10.1126/science.1241475

Thal DM et al (2016) Crystal structures of the M1 and M4 muscarinic acetylcholine receptors. Nature 531(7594):335–340. https://doi.org/10.1038/nature17188

Toyoda Y et al (2019) Ligand binding to human prostaglandin E receptor EP4 at the lipid-bilayer interface. Nat Chem Biol 15(1):18–26. https://doi.org/10.1038/s41589-018-0131-3

van Koppen CJ, Kaiser B (2003) Regulation of muscarinic acetylcholine receptor signaling. Pharmacol Ther 98(2):197–220. https://doi.org/10.1016/s0163-7258(03)00032-9

Violin JD, Lefkowitz RJ (2007) Beta-arrestin-biased ligands at seven-transmembrane receptors. Trends Pharmacol Sci 28(8):416–422. https://doi.org/10.1016/j.tips.2007.06.006

Vuckovic Z et al (2019) Crystal structure of the M5 muscarinic acetylcholine receptor. Proc Natl Acad Sci U S A 116(51):26001–26007. https://doi.org/10.1073/pnas.1914446116

Wacker D et al (2013) Structural features for functional selectivity at serotonin receptors. Science 340(6132):615–619. https://doi.org/10.1126/science.1232808

Waltenspühl Y et al (2020) Crystal structure of the human oxytocin receptor. Sci Adv 6(29):eabb5419. https://doi.org/10.1126/sciadv.abb5419

Wang C et al (2013) Structural basis for molecular recognition at serotonin receptors. Science 340(6132):610–614. https://doi.org/10.1126/science.1232807

Wang S et al (2017) D_4 dopamine receptor high-resolution structures enable the discovery of selective agonists. Science 358(6361):381–386. https://doi.org/10.1126/science.aan5468

Wang L et al (2018) Structures of the human PGD_2 receptor $CRTH_2$ reveal novel mechanisms for ligand recognition. Mol Cell 72(1):48–59 e4. https://doi.org/10.1016/j.molcel.2018.08.009

Warne T et al (2008) Structure of a β1-adrenergic G-protein-coupled receptor. Nature 454(7203):486–491. https://doi.org/10.1038/nature07101

Wasilko DJ et al (2020) Structural basis for chemokine receptor CCR_6 activation by the endogenous protein ligand CCL_{20}. Nat Commun 11(1):3031. https://doi.org/10.1038/s41467-020-16820-6

Whalen EJ, Rajagopal S, Lefkowitz RJ (2010) Therapeutic potential of β-arrestin- and G protein-biased agonists. Trends Mol Med 17(3):126–139. https://doi.org/10.1016/j.molmed.2010.11.004

Wu B et al (2010) Structures of the $CXCR_4$ chemokine GPCR with small-molecule and cyclic peptide antagonists. Science 330(6007):1066–1071. https://doi.org/10.1126/science.1194396

Xu F et al (2011) Structure of an agonist-bound human A2A adenosine receptor. Science 332(6027):322–327. https://doi.org/10.1126/science.1202793

Xu J et al (2019) Conformational complexity and dynamics in a muscarinic receptor revealed by NMR spectroscopy. Mol Cell 75(1):53-65.e7. https://doi.org/10.1016/j.molcel.2019.04.028

Yan W et al (2020) Structure of the human gonadotropin-releasing hormone receptor GnRH1R reveals an unusual ligand binding mode. Nat Commun 11(1):5287. https://doi.org/10.1038/s41 467-020-19109-w

Yang F et al (2015) Phospho-selective mechanisms of arrestin conformations and functions revealed by unnatural amino acid incorporation and 19F-NMR. Nat Commun 6:8202. https://doi.org/10. 1038/ncomms9202

Yang Z et al (2018) Structural basis of ligand binding modes at the neuropeptide Y Y_1 receptor. Nature 556(7702):520–524. https://doi.org/10.1038/s41586-018-0046-x

Yin J et al (2015) Crystal structure of the human OX_2 orexin receptor bound to the insomnia drug suvorexant. Nature 519(7542):247–250. https://doi.org/10.1038/nature14035

Yin J et al (2016) Structure and ligand-binding mechanism of the human OX_1 and OX_2 orexin receptors. Nat Struct Mol Biol 23(4):293–299. https://doi.org/10.1038/nsmb.3183

Yu J et al (2020) Determination of the melanocortin-4 receptor structure identifies Ca^{2+} as a cofactor for ligand binding. Science 368(6489):428–433. https://doi.org/10.1126/science.aaz8995

Yuan D et al (2020) Activation of the α_{2B} adrenoceptor by the sedative sympatholytic dexmedeto-midine. Nat Chem Biol 16(5):507–512. https://doi.org/10.1038/s41589-020-0492-2

Zhang C et al (2012) High-resolution crystal structure of human protease-activated receptor 1. Nature 492(7429):387–392. https://doi.org/10.1038/nature11701

Zhang J et al (2014) Agonist-bound structure of the human $P2Y_{12}$ receptor. Nature 509(7498):119–122. https://doi.org/10.1038/nature13288

Zhang H et al (2015a) Structural basis for ligand recognition and functional selectivity at angiotensin receptor. J Biol Chem 290(49):29127–29139. https://doi.org/10.1074/jbc.M115.689000

Zhang D et al (2015b) Two disparate ligand-binding sites in the human $P2Y_1$ receptor. Nature 520(7547):317–321. https://doi.org/10.1038/nature14287

Zou X et al (2018) Investigating the effect of two-point surface attachment on enzyme stability and activity. J Am Chem Soc 140(48):16560–16569. https://doi.org/10.1021/jacs.8b08138

Chapter 2
Purification of G Protein-Coupled Receptors

Abstract Apart from the singular exception of rhodopsin, GPCRs are naturally poorly abundant in the human body. As such, producing milligram amounts of purified, active, and stable GPCRs is not trivial and has long been considered challenging work to realize. With persistent technological and methodological efforts, a growing number of recombinant and active GPCRs has been successfully purified. In this chapter, we aim to discuss the solubilization and stabilization of GPCR using different detergents and styrene-maleic acid–lipid particles. We also highlight the commonly used purification techniques for GPCRs in literature. Finally, the heterologous expression systems for GPCRs, ranging from bacteria and yeasts to insects, mammalian cells, and *Drosophila melanogaster*, and the successful purification cases are reviewed.

Keywords GPCR purification · Solubilization · Stabilization · Heterologous expression

Abbreviations

GPCRs	G-protein coupled receptors
CMCs	Critical micelle concentrations
HLB	Hydrophile-lipophile balance
DM	n-Dodecyl-β-D-maltoside
DDM	n-Dodecyl-β-D-maltopyranoside
LMNG	Lauryl maltose neopentyl glycol
CHS	Cholesteryl hemisuccinate
CHAPS	3-[(3-Cholamidopropyl)dimethylammonio]-1-propanesulfonate
SMA	Styrene-co-maleic acid
SMALPs	Styrene-co-maleic acid lipid particles
DIBMA	Diisobutylene maleic acid copolymer
SMI	Styrene-N-phenylmaleimide copolymer
PMA	Acrylic acid and maleic acid copolymer
Twin-Strep	$10 \times$ His

X. Zhao et al., *G Protein-Coupled Receptors*, SpringerBriefs in Molecular Science,
https://doi.org/10.1007/978-981-99-0078-7_2

Strep   6 × His
IEX   Ion exchange chromatography
GF   Gel filtration
HIC   Hydrophobic interaction
RPC   Reversed phase chromatography
GFP   Green fluorescent protein
MOI   Multiplicity of infection
PRCs   Photoreceptor cells

2.1 GPCR Solubilization and Stabilization

Like other transmembrane proteins, GPCRs are embedded in the lipid bilayer. This makes the protein surrounded by two discrete Physico-chemical environments simultaneously, including the aqueous phase in combination with charged lipid headgroups of the membrane and the hydrophobic membrane interior. The environments contributed to the fold of GPCRs facilitating the function of the protein. Problems are introduced when experiments require the GPCRs to be extracted from the membrane environment for purification purposes. An essential step towards purification is the solubilization of GPCRs, i.e., using suitable amphiphilic detergents to remove receptors from the native membrane environment and disperse them homogeneously (Chattopadhyay et al. 2015).

2.1.1 Detergents

Detergents often self-assemble to form noncovalent aggregates, which are thermodynamically stable micelles. The formation of micelles is related to the critical micelle concentrations (CMCs) of detergents, typically in the range of millimolar, which is crucial for the solubilization and reconstitution of GPCRs (Fig. 2.1). Typically, the loss of receptor function happens when detergents are used at concentrations above their CMC. Another parameter, namely hydrophile-lipophile balance (HLB), which can be measured as the weight percentage of hydrophilic versus lipophilic groups in a detergent, is highly related to the ability of a detergent to solubilize GPCRs. The value of HLB in the range of 12–20 is recommended for efficient solubilization of GPCRs without denaturation.

The detergent used to solubilize GPCRs can be classified into three categories: (i) the nonionic detergent; (ii) the zwitterionic detergent; and (iii) the anionic detergent. Harshness is another indicator to describe the efficacy of disrupting intra- and intermolecular interactions. Overall, ionic detergents are considered harsh, zwitterions are milder and non-ionic detergents are regarded as mild. Nonionic detergents like n-dodecyl-β-D-maltoside (DM, w/v = 1.5–2%), n-dodecyl-β-D-maltopyranoside

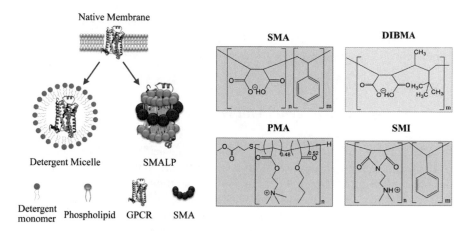

Fig. 2.1 Structures of different detergents

(DDM, w/v = 0.5–1.5%), and lauryl maltose neopentyl glycol (LMNG, w/v = 1%) have become popular for effectively solubilizing GPCRs (~45%) and can be regarded as an evidence-based starting point (Wiseman et al. 2020). DDM and LMNG are mild detergents and form large micelles, which offer the advantage of receptor aggregation. However, the low CMC concentrations (DDM: 0.0087%; LMNG: 0.001%) and large micelle sizes (DDM: 72 kDa; LMNG: >90 kDa) of the two detergents make it hard to exchange them for another detergent. DM is a harsher detergent with higher CMC (0.1%) and a smaller micelle size (33 kDa). It is easier to exchange DM, but receptor stability and solubilization efficiency need to be taken into account when it is applied. Cholesteryl hemisuccinate (CHS) is often used as an additive to DDM/LMNG in a concentration range of 0.1–0.2% (w/v). It often induces a bicelle-like architecture and offers a more stable environment for GPCRs. However, the addition of CHS will increase the micelle size and interfere with the binding of affinity tags to purification resins. To overcome this problem, the concentration of CHS is often decreased by 10–100 times in purification buffers. Zwitterionic detergent like 3-[(3-cholamidopropyl)dimethylammonio]-1-propanesulfonate (CHAPS) is also popular to solubilize GPCRs. It has the characteristics of the N-alkyl sulfobetaine-type polar group and the bile salt hydrophobic group, thus, more efficient in solubilizing GPCRs than cholate. Notably, CHAPS lacks a circular dichroic signal in the far-UV region and has low absorbance at 280 nm, making it ideal for optical spectroscopic studies of GPCRs.

Even though some GPCRs may successfully solubilize and be sufficiently stable in detergents, the formed micelle structures are poor mimics of a lipid bilayer environment, which is important in modulating GPCR activity. The active/inactive of GPCRs may be influenced by the lipids surrounding them, which provide lateral pressure to the receptors. Micelles formed by detergents do not offer the same lateral pressure and the directly bound lipids required for receptor stability/function may be removed. It is preferential to develop new techniques to prevent receptor destabilization/inactivation during solubilization. In most cases, detergents should be chosen empirically and are considered as a receptor-demanding approach.

2.1.2 Styrene Maleic Acid Lipid Particles

To solve the above problems, poly (styrene-co-maleic acid) (SMA) was introduced to incorporate into cell membranes (Fig. 2.1). SMA can spontaneously assemble into nano-scale (~10 nm) SMA lipid particles (SMALPs), effectively act like a 'molecular pastry cutter', and possess the expected thickness of a cell membrane. By using SMA, GPCRs can be rapidly and effectively purified in a detergent-free manner. Notably, $A_{2A}R$-SMALPs were stable in more than 5 freeze–thaw cycles without reduction in the receptor-ligand binding efficiency and retained 75% activity of the receptor after 15 days of storage at 4 °C. However, the binding efficiency declined to 0% by day 3 when solubilized the receptor in detergent (Jamshad et al. 2015). The remarkable stability of $A_{2A}R$ makes SMALPs more applicable and flexible than typical detergents. Due to the presence of styrene groups, SMA has strong absorbance in the far-UV region. It also has a limited operating pH range to solubilize cell membranes and needs to be precipitated at low millimolar concentrations of divalent cations (Lavington and Watts 2020). Some alternative polymers were also developed to form lipid nanoparticles, such as diisobutylene maleic acid copolymer (DIBMA), styrene-N-phenylmaleimide copolymer (SMI), and acrylic acid and maleic acid copolymer (PMA). DIBMA and SMI are less sensitive to cation-induced precipitation and PMA has a different scaffold. All of them can assemble into discoidal bilayers with diameters of 10–20 nm.

2.2 Purification Techniques for GPCRs

Chromatographic methods are the mainstream way to facilitate GPCR purification. GPCRs are often engineered to have peptide tags, like poly-histidines, strep tags, and FLAG tag, or protein domains, such as maltose binding protein and thioredoxin, which allow the interactions between the tags and their specific affinity matrices in the presence of detergents. The tags can be inserted either on the N-terminal, C-terminal, or even on extra-membranous loops of GPCRs.

Peptide tags can be used in affinity chromatography relying on immobilized metals, ligands, or antibodies. GPCRs with poly-histidine tags can be purified by immobilized metal-ion chromatography (e.g. Ni^{2+}- and Co^{2+}-coated resins), which is also considered as the commonly used first chromatographic step in GPCR purification. Compared with the nickel resin, cobalt has a lower affinity for histidine, thus helping to remove non-specific proteins and enrich the receptors at a very low expression level. The imidazole ring-containing ligands may prevent the binding of his-tagged receptors (e.g. histamine receptor) to Ni^{2+}-coated resins. As an alternative, GPCRs can be engineered using a Strep-tag (NH_2-WSHPQFEK-COOH). Biotin-coated resins are often used to separate a Strep-tagged GPCR and the receptors can be eluted under gentle, physiological conditions with high purity in a one-step fashion (Jazayeri et al. 2017). It is noted that the partial burial of the poly-histadine or Strep tags in the detergent micelles may affect the purification efficiency of the GPCRs. Improvements in the purification yield can be achieved using $10 \times$ His (Twin-Strep) rather than $6 \times$ His (Strep) in combination with the insertion of a flexible peptide linker and protease cleavage site (Errey and Fiez-Vandal 2020). Another advantage of using these tags is their abilities to be served as a measure of GPCR expression, solubilization, and purification. Since the antibodies of the tags are commercially available and can give reliable readout. Although many antibodies that recognize GPCRs are commercially available, some of them lack enough specificity to be widely used.

It is advantageous to use a fusion system by affinity chromatography to realize the GPCR purification at a desired level of purity. Although the peptide or protein domain tags make protein purification easier, their presence may alter the GPCRs' function: (i) N-terminal tags may affect ligand binding efficiency; (ii) C-terminal tags may avoid the G protein activation process. Therefore, fusion tags are often removed after the purification step and this is commonly achieved by the choice of desirable protease. When the heterologous expression of the GPCR of interest, the protease cleavage site can be inserted between the amino acid sequences of the affinity tags and GPCRs. It should be noted that the activity of protease may be inhibited by the detergent used in GPCR solubilization and purification. Thus, the detergent exchange is needed to ensure the effective cleavage of the fusion tags by proteases.

2.3   GPCR Purification from Native Tissues

It is known that most of the GPCRs have relatively low expression in native tissues without any fusion tags. This brings more challenges in the purification of GPCRs from native tissues. A common way is to use ligand-guided affinity chromatography for purification. The resin with covalently immobilized specific ligands of a GPCR is now available for the purification of the receptor. The ligands like alprenolol (Søren et al. 2011), neurotensin (White et al. 2004), and aminobenztropine (Haga et al. 2012) have been successfully modified on resin surfaces and used for the purification of β_2-AR, NTS_1R and M_2R, respectively. The merit of the ligand-guided purification method is that only functional GPCRs will be purified. This is advantageous when the active and inactive GPCRs need to be separated. However, this method often suffered from low binding capacity and poor elution efficiency.

However, if there is no suitable ligand for purification so that affinity chromatography cannot be used, or a higher degree of purity is required, a multi-step purification will be necessary. There are many chromatographic techniques available for GPCRs purifications that separate according to differences in specific protein properties: (i) ion exchange chromatography (IEX) uses the charge difference in the protein surface to realize separation; (ii) gel filtration (GF) can achieve separation via the size of the proteins; (iii) hydrophobic interaction (HIC) and reversed-phase chromatography (RPC) separate proteins through their hydrophobicity difference. A standard protocol is to use IEX-HIC-GF combination techniques if high purity levels are required, i.e., IEX for capture, HIC for intermediate purification, and GF for the final purification.

2.4 Heterologous Expression of GPCRs

2.4.1 GPCRs Expressed in E. coli

E. coli is the most attractive host for the overexpression of the heterologous protein. It has become a laboratory workhorse for multiple reasons: (i) *E. coli* is easy to be developed as an effective genetic tool by modifying the strains and plasmids, which allows for high levels of protein expression; (ii) the strain can grow very fast through tunning the fermentation conditions to produce the yield of the target proteins; (iii) the costs to culture *E. coli* is relatively lower than other hosts, which can be affordable by most of the researchers.

Compared with other expression systems, the overproduction success of GPCRs in *E. coli* is often hindered by the limited native processing ability of prokaryotic organisms for the inherent defect in protein trafficking and folding machinery, essential protein folding (e.g., disulfide bond formation), and post-translational modifications (e.g., glycosylation). A high-level expression may lead to the misfolding and aggregation of GPCRs and form inclusion bodies. To overcome this issue, proteins will be expressed at lower temperatures or strain selection will be needed. Another way to optimize experimental conditions and monitor the right-folded GPCR expression is to fuse green fluorescent protein (GFP) to the C-terminal end of the proteins since only properly folded fusion proteins would display GFP fluorescence (Drew et al. 2006). Besides, other fusion partners and selective mutagenesis are also used to enhance the correct folding, increase the overall stability, and improve the success rate of GPCRs (Zhao et al. 2016; Yamamoto et al. 2022). In addition, the lipid membrane component of *E. coli* is different from eukaryotic cells, which will affect the activity of lipid-dependent GPCR (Goddard et al. 2013). Despite the above limitations, the use of *E. coli* for GPCR production is benefic for downstream applications like NMR analysis. It is easy to conduct isotopic labeling for GPCRs, which is convenient for NMR studies (Marina et al. 2018). Furthermore, the genetic amenability of *E. coli* provides possibilities to select GPCR variants with enhanced expression, stability, and functionality. Nowadays, a number of functional GPCRs have been successfully expressed in *E. coli*, such as NTS_1R (Bumbak 2019), D_2R (Boritzki et al. 2022), turkey β_1-AR (Abiko et al. 2020), CB_2R (Laffitte et al. 2022), human

TAS$_1$R2 (Laffitte et al. 2022), and human β_2-AR. In spite of the prokaryotic nature, *E. coli* has clear merit for a subset of GPCR production.

2.4.2 GPCRs Expressed in Yeast

Yeast expression systems are another attractive host for GPCR expression. It combines the advantages of prokaryotic (e.g. robust DNA repair and recombination machinery, easy culture, and high-density inexpensive growth) and eukaryotic systems (e.g. post-translational modification machinery). These characteristics make yeast an important tool to investigate the stability, structure, and signaling of GPCRs (Wiseman et al. 2020).

Saccharomyces cerevisiae (*S. cerevisiae*) and *Pichia pastoris* (*P. pastoris*) are the two most important yeast species used as hosts for exogenous protein expression (Li et al. 2022). *S. cerevisiae* has only two signaling pathways, one affects glucose sensing and the other is involved in mating (Liu et al. 2016). The first heterologous GPCR to be coupled to the mating pathway in *S. cerevisiae* was the human β_2-AR (King et al. 1990). It provides a "null" background when studying human GPCRs. Since then, a subset of human GPCRs has been successfully expressed in this system for protein activation and signaling, as well as the agonist and antagonist identification studies (Liu et al. 2016). High yields of functional receptors have been expressed in *S. cerevisiae*, including A$_1$R, A$_{2B}$R, A$_{2A}$R, SSTR5, mAChRs, NTSR1, AT$_1$R, CXCR4, C5aR, GLP-1R, HTR$_{1A}$. The rat voltage-dependent potassium ion channel was the first recombinant mammalian membrane protein expressed in *P. pastoris* for crystallization (Long et al. 2005). Recently, the human vitamin K epoxide reductase and the proton activated chloride channel TMEM206 were produced in *P. pastoris* for X-ray crystallography and cryo-electron microscopy analyses (Liu et al. 2021; Deng et al. 2021). To benefit from the best attributes of the two hosts, *P. pastoris* is often used as an initial expression host since it offers stable integration of the expression plasmids. If the protein expression yields are low, *S. cerevisiae* can be applied for troubleshooting because this species uses episomal plasmids and results in high cell densities on glycerol (Ayub et al. 2022).

It is noted that cholesterol is a crucial component in mammalian cell membranes to stabilize and maintain the correct function of GPCRs (Paila et al. 2008). Similar to *E. coli*, yeast membranes do not have cholesterol. Cholesterol can allosterically modulate GPCRs, like chemokine receptor CCR3 and cannabinoid receptor CB$_2$ (van Aalst and Wylie 2021; Yeliseev et al. 2021), its replacement with ergosterol in yeast expression systems may lead to the inactive of the receptors (Wiseman et al. 2020). To solve this problem, a new *P. pastoris* has been engineered and manipulated to synthesize cholesterol, which is beneficial for GPCR expression (Hirzb et al. 2013).

2.4.3   GPCRs Expressed in Insect Cell-Line

The insect cells are the most widely used expression system to obtain milligram quantities of GPCRs (Boivineau et al. 2020). During the last 10 years, approximately 31% of the eukaryotic membrane proteins were obtained from the organism for structural studies (Kesidis et al. 2020). Expression of heterogenous GPCRs is achieved via infection by *Autographa californica* baculovirus, which is a double-stranded, multiple nuclear polyhedrosis virus, surrounded by a lipid membrane. The majority of GPCR researches utilize the *Spodoptera frugiperda* (*Sf*9 and *Sf*21) cell line, in preference to *Trichoplusia ni* (His$_5$ and MG$_I$) cells (Sarramegna et al. 2003). Screening of the optimized cell line is essential since the protein expression levels can vary among different cell lines. The insect cells can produce high yield of proteins and have the ability for post-translational modifications. They commonly grow in serum-free shaker cultures, which decreases costs and enables relatively easy scaleup compared with the GPCRs expressed in mammalian cells.

There are two main systems often used to construct the recombinant baculovirus, Bac-to-Bac™ and flashBAC™ systems (Fig. 2.2). Three steps are needed in Bac-to-Bac™ system: (i) the GPCR gene of interest is first cloned into a pFastBac™ vector, which is under controlled by a polyhedrin promoter; (ii) the pFastBac™ vector is subsequent transformed into DH10Bac™ *E. coli* cells, which have a bacmid and a helper plasmid. The bacmid is composed of the baculovirus genome and a transposon. The presence of transposon makes the gene of the interest transferred from the pFastBac™ vector to the bacmid; (iii) then the recombinant bacmid is isolated and purified to transfect the insect cells, leading to the production and release of the baculovirus into the culture medium (Vaitsopoulou et al. 2022). In the flashBAC™ system, the baculovirus genome lacks ORF1629, thus, the virus cannot replicate in insect cells. Then the flashBAC™ circular DNA is introduced, of which the polyhedrin coding gene is replaced by a bacterial artificial chromosome (BAC). Transfer vectors like pOET contain the full ORF1629 gene and could incorporate the gene of GPCR. The flashBAC™ DNA and the transfer vector can be co-transfected into insect cells, resulting in the recovery of the ORF1629 function. In the meantime, the GPCR gene replaces the BAC sequence, leading to the replication of the viral DNA. The last step is to harvest the recombinant virus from the culture medium and amplified for protein production. The flashBAC™ is simpler than the Bac-to-Bac™ system and no recombinant bacmid is needed (Vaitsopoulou et al. 2022).

The insect cells present relatively long generation times (~24 h), which is much longer than the *E. coli* expression system (~20 min). It is noted that the plasmid containing heterogenous GPCR genes and viral DNA need to be co-transfected into the insect cells, allowing the insertion of the GPCR genes into the viral genome. The polyhedrin promoter is a 'late' promoter and activates 8–24 h after the infection process (Saarenpää et al. 2015). Therefore, receptor production happens at the end of the cell life; it is beneficial for the GPCRs, which are toxic to the cells. In addition, to generate the recombinant baculovirus, titration is needed to realize an optimal multiplicity of infection (MOI) (Islam et al. 2021). Directly quantifying the virus

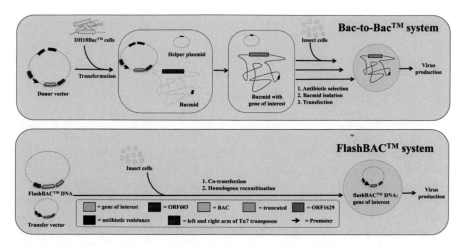

Fig. 2.2 Comparison of the Bac-to-Bac™ and flashBAC™ system for virus production

is very difficult and an excess of the virus will result in cell death before they can be harvested. Moreover, the lipid composition of insect cells is also different from mammalian cells. Insect cell membranes have a high content of phosphatidylinositol, are low in cholesterol, and have no phosphatidylserine (Milić and Veprintsev 2015). Overall, insect cells are the key approach for GPCR overexpression.

2.4.4 GPCRs Expressed in Mammalian Cell-Line

Mammalian cell lines are often preferred as expression hosts since the cells provide the endogenous machinery for post-translational modification and adequate trafficking, making them ideal for GPCR function and pharmacology characterization (Burnat et al. 2020). It is demonstrated that post-translational modification may be prejudicious for crystal formation in structure biological study. For instance, the flexibility and heterogeneity of glycan residues in the glycosylation site may affect the formation of ordered crystals (Milić and Veprintsev 2015). This issue can be avoided by mutation of the N-glycosylation sites if the conformation of the receptor is still stable. Another choice is to use the Nacetylglucosaminyl transferase I-deficient cell line to control the glycans (Reeves 2021). Except for the glycosylation, phosphorylation and palmitoylation of GPCRs can be also achieved by the enzymes in mammalian cells, which is important for receptor pharmacology (Thibeault and Ramachandran 2020). Moreover, the membrane lipid composition of the mammalian cell line is similar to the endogenous GPCR, which is a prerequisite for ensuring the function of GPCRs. Overall, mammalian cell lines are the ideal hosts for human GPCRs. However, deficiencies also exist, which included the low yields and high costs of

GPCR expression. Considering the high costs, the hosts for GPCR expression should be carefully selected.

2.4.5 GPCRs Expressed in Drosophila melanogaster

The fruit fly, *Drosophila melanogaster*, is an alternative to express exogenous GPCRs developed in recent years. The eyes of the fruit fly have unique architecture containing photoreceptor cells (PRCs). The PRCs have rhabdomeres, a kind of membrane stacks, providing adequate surface area for the expression and folding of GPCRs (Panneels et al. 2011). To realize heterologous GPCR expression, GAL4-UAS system was developed (Phelps and Brand 1998). The yeast GAL4 protein is a transcription factor, which facilitates the expression of target protein through binding to an Upstream Activating Sequence (UAS). To obtain the heterologous GPCR, the driver-strain containing GAL4 and UAS strain maintaining the transgene of interest are mated together. The resulting offspring will express the target GPCR in PRCs. The costs of culturing the flies are cheap and the experiments do not need to perform in sterile conditions, however, this method requires access to fly genetics expertise and some post-translational modifications are also different from the human body (Wiseman et al. 2020). Despite this, a number of GPCRs have been successfully expressed in *Drosophila melanogaster* like metabotropic glutamate receptor (Eroglu et al. 2002), human serotonin 5-HT$_7$ receptor (Courant et al. 2022), ocular albinism type 1 (Sun et al. 2019).

Collectively, producing and purifying milligram amounts of active and stable GPCRs is not trivial. There is still a long way to go to achieve high purity and active GPCRs.

References

Abiko LA et al (2020) Efficient production of a functional G protein-coupled receptor in E. coli for structural studies. J Biomol NMR 75(1):25–38. https://doi.org/10.1007/s10858-020-00354-6

Ayub H et al (2022) Membrane protein production in the yeast P. pastoris. In: Mus-Veteau I (ed) Heterologous expression of membrane proteins. Methods in molecular biology, vol 2507. Humana, New York, NY, pp 187–199. https://doi.org/10.1007/978-1-0716-2368-8_10

Boivineau J, Haffke M, Jaakola VP (2020) Membrane protein expression in insect cells using the baculovirus expression vector system. In: Perez C (ed) Expression purification and structural biology of membrane proteins. Methods Mol Biol, vol 2127. Humana, New York, NY, pp 63–80. https://doi.org/10.1007/978-1-0716-0373-4_5

Boritzki V et al (2022) Optimizing the expression of human dopamine receptors in Escherichia coli. Int J Mol Sci 22(16):8647. https://doi.org/10.3390/ijms22168647

Bumbak F et al (2019) Expression and purification of a functional E. coli ^{13}CH$_3$-methionine-labeled thermostable neurotensin receptor 1 variant for solution NMR studies. In: Tiberi M (ed) G protein-coupled receptor signaling. Methods in molecular biology, vol 1947. Humana Press, New York, NY, pp 31–35. https://doi.org/10.1007/978-1-4939-9121-1_3

Burnat G et al (2020) The functional cooperation of 5-HT1A and mGlu4R in HEK-293 cell line. Pharmacol Rep 72(5):1358–1369. https://doi.org/10.1007/s43440-020-00114-1

Chattopadhyay A, Rao BD, Jafurulla M (2015) Solubilization of G protein-coupled receptors: a convenient strategy to explore lipid–receptor interaction. In: Shukla AK (ed) Methods in enzymology, vol 557. Academic Press, pp 117–134. https://doi.org/10.1016/bs.mie.2015.01.001

Courant F et al (2022) Expression of the human serotonin 5-HT7 receptor rescues phenotype profile and restores dysregulated biomarkers in a drosophila melanogaster glioma model. Cells 11:1281. https://doi.org/10.3390/cells11081281

Deng Z et al (2021) Cryo-EM structure of a proton-activated chloride channel TMEM206. Sci Adv 7:eabe5983. https://doi.org/10.1126/sciadv.abe5983

Drew D et al (2006) Optimization of membrane protein overexpression and purification using GFP fusions. Nat Methods 3:303–313. https://doi.org/10.1038/nmeth0406-303

Eroglu C et al (2002) Functional reconstitution of purified metabotropic glutamate receptor expressed in the fly eye. Embo Rep 3:491–496. https://doi.org/10.1093/embo-reports/kvf088

Errey CJ, Fiez-Vandal C (2020) Production of membrane proteins in industry: the example of GPCRs. Protein Expres Purif 169:105569. https://doi.org/10.1016/j.pep.2020.105569

Goddard AD et al (2013) Lipid-dependent GPCR dimerization. In Conn PM (ed) Methods in cell biology, vol 117. Academic Press, pp 341–357. https://doi.org/10.1016/B978-0-12-408143-7.000 18-9

Haga K et al (2012) Structure of the human M2 muscarinic acetylcholine receptor bound to an antagonist. Nature 482:547–551. https://doi.org/10.1038/nature10753

Hirzb M et al (2013) A novel cholesterol-producing Pichia pastoris strain is an ideal host for functional expression of human Na, K-ATPase $\alpha3\beta1$ isoform. Appl Microbiol Biote 97:9465–9478. https://doi.org/10.1007/s00253-013-5156-7

Islam S et al (2021) C-C Chemokine receptor-like 2 (CCRL2) acts as coreceptor for human immunodeficiency virus-2. Brief Bioinform 22:4. https://doi.org/10.1093/bib/bbaa333

Jamshad M et al (2015) G-protein coupled receptor solubilization and purification for biophysical analysis and functional studies, in the total absence of detergent. Biosci Rep 35(2):e00188. https://doi.org/10.1042/BSR20140171

Jazayeri A et al (2017) Crystal structure of the GLP-1 receptor bound to a peptide agonist. Nature 546:254–258. https://doi.org/10.1038/nature22800

Kesidis A et al (2020) Expression of eukaryotic membrane proteins in eukaryotic and prokaryotic hosts. Methods 180:3–18. https://doi.org/10.1016/j.ymeth.2020.06.006

King K et al (1990) Control of yeast mating signal transduction by a mammalian beta 2-adrenergic receptor and Gs alpha subunit. Science 250:121–123. https://doi.org/10.1126/science.2171146

Laffitte A et al (2022) Functional characterization of the Venus flytrap domain of the human TAS1R2 sweet taste receptor. Int J Mol Sci 23:9216. https://doi.org/10.3390/ijms23169216

Lavington S, Watts A (2020) Lipid nanoparticle technologies for the study of G protein-coupled receptors in lipid environments. Biophysical Rev 12:1287–1302. https://doi.org/10.1007/s12551-020-00775-5

Li X et al (2022) Heterologous expression and purification of GPCRs. Methods Mol Biol 2507:295–312. https://doi.org/10.1007/978-1-0716-2368-8_15

Liu R et al (2016) Human G protein-coupled receptor studies in Saccharomyces cerevisiae. Biochem Pharmacol 114:103–115. https://doi.org/10.1016/j.bcp.2016.02.010

Liu S et al (2021) Structural basis of antagonizing the vitamin K catalytic cycle for anticoagulation. Science 371:eabc5667. https://doi.org/10.1126/science.abc5667

Long SB et al (2005) Crystal structure of a mammalian voltage-dependent shaker family K+ channel. Science 309:897–903. https://doi.org/10.1126/science.1116269

Marina C et al (2018) Illuminating the energy landscape of GPCRs: the key contribution of solution-state NMR associated with Escherichia coli as an expression host. J Am Chem Soc 57:2297–2307. https://doi.org/10.1021/acs.biochem.8b00035

Milić D, Veprintsev DB (2015) Large-scale production and protein engineering of G protein-coupled receptors for structural studies. Front Pharmacol 6:66. https://doi.org/10.3389/fphar.2015.00066

Paila YD, Tiwari S, Chattopadhyay A (2008) Are specific nonannular cholesterol binding sites present in G-protein coupled receptors? Biochim Biophys Acta 1788(2):295–302. https://doi.org/10.1016/j.bbamem.2008.11.020

Panneels V et al (2011) Drosophila photoreceptor cells exploited for the production of eukaryotic membrane proteins: receptors, transporters and channels. PLoS ONE 6:e18478. https://doi.org/10.1371/journal.pone.0018478

Phelps CB, Brand AH (1998) Ectopic gene expression in Drosophila using GAL4 system. Methods 14:367–379. https://doi.org/10.1006/meth.1998.0592

Reeves PJ (2021) Construction of recombinant cell lines for GPCR expression. In: Martins (ed) G protein-coupled receptor screening assays. Methods in molecular biology, vol 2268. Humana, New York, NY, pp 43–60. https://doi.org/10.1007/978-1-0716-1221-7_3

Saarenpää T et al (2015) Chapter nine—Baculovirus-mediated expression of GPCRs in insect cells. In: Shukla AK (ed) Methods in enzymol, vol 556. Academic Press, pp 185–218. https://doi.org/10.1016/bs.mie.2014.12.033

Sarramegna V et al (2003) Heterologous expression of G-protein-coupled receptors: comparison of expression systems from the standpoint of large-scale production and purification. CMLS Cell Mol Life Sci 60:1529–1546. https://doi.org/10.1007/s00018-003-3168-7

Søren GFR et al (2011) Crystal structure of the β2 adrenergic receptor–Gs protein complex. Nature 477:549–555. https://doi.org/10.1038/nature10361

Sun L et al (2019) Ocular albinism type 1 regulates deltamethrin tolerance in lymantria dispar and drosophila melanogaster. Front Physiol 10:766. https://doi.org/10.3389/fphys.2019.00766

Thibeault PE, Ramachandran R (2020) Role of the helix-8 and C-terminal tail in regulating proteinase activated receptor 2 signaling. ACS Pharmacol Transl Sci 3:868–882. https://doi.org/10.1021/acsptsci.0c00039

Vaitsopoulou A et al (2022) Membrane protein production in insect cells. In: Mus-Veteau I (ed) Heterologous expression of membrane proteins. Methods in molecular biology, vol 2507. Humana, New York, NY, pp 223–240. https://doi.org/10.1007/978-1-0716-2368-8_12

van Aalst E, Wylie BJ (2021) Cholesterol is a dose-dependent positive allosteric modulator of CCR3 ligand affinity and G protein coupling. Front Mol Biosci 8:724603. https://doi.org/10.3389/fmolb.2021.724603

White JF et al (2004) Automated large-scale purification of a G protein-coupled receptor for neurotensin. FEBS Lett 564:289–293. https://doi.org/10.1016/S0014-5793(04)00195-4

Wiseman DN et al (2020) Expression and purification of recombinant G protein-coupled receptors: a review. Protein Expres Purif 167:105524. https://doi.org/10.1016/j.pep.2019.105524

Yamamoto T et al (2022) A methodology for creating mutants of G-protein coupled receptors stabilized in active state by combining statistical thermodynamics and evolutionary molecular engineering. Protein Sci 31:10. https://doi.org/10.1002/pro.4425

Yeliseev A et al (2021) Cholesterol as a modulator of cannabinoid receptor CB2 signaling. Sci Rep 11:3706. https://doi.org/10.1038/s41598-021-83245-6

Zhao JY et al (2016) Complete genome sequence of defluviimonas alba cai42T, a microbial exopolysaccharides producer. J Biotechnol 239:9–12. https://doi.org/10.1016/j.jbiotec.2016.09.017

Chapter 3
Oriented Immobilization of G Protein-Coupled Receptors

Abstract GPCR immobilization is fundamental for many bioanalytical methods, which are crucial for receptor-ligand interaction analysis and drug discovery. A desirable way to immobilize GPCRs on solid surfaces is oriented immobilization. In this way, the receptors can be optimal in their orientation, conformation, and activity or functional studies. In this chapter, we review the affinity-based non-covalent and site-specific covalent immobilization methods developed for capturing GPCR on solid surfaces. The *cons* and *pros* of each method are also discussed.

Keywords Affinity-based non-covalent immobilization · Site-specific covalent immobilization · Bioorthogonal chemistry · Biologically mediated site-specific immobilization

Abbreviations

GPCRs	G protein-coupled receptors
NHS	N-Hydroxysuccinimide
5-HT$_3$R	5-Hydroxytryptamlne receptor
nAChR	Nicotinic acetylcholine receptor
EDTA	Ethylenediaminetetraacetic acid
NTS-1	Neurotensin receptor-1
β_2-AR	β_2-Adrenergic receptor
ssDNA	Single-stranded DNA
EGFR	Epidermal growth factor receptor
SrtA	Sortase A
hAGT	O^6-Alkylguanine-DNA alkyltransferase
BG	O^6-Benzylguanine
Halo	Haloalkane dehalogenase
EPL	Expressed protein ligation
MNPs	Magnetic nanoparticles
CalBP	Calmodulin-binding peptide
CBD	Cellulose-binding domains

MBP   Maltose-binding protein
Im7   Immunity protein 7
GST   Glutathione S-transferase
Pr-A   Protein A
HA   Hemagglutinin A
VSV-G   Vesicular stomatitis virus glycoprotein

3.1   Affinity-Based Non-Covalent Immobilization

In most cases, protein immobilization can be achieved via either random or site-specific attachment that has been extensively employed in protein labeling. Chemical strategies relying on both noncovalent and covalent reactions between certain supports and the functional groups of proteins have the capacity to generate random immobilization. Noncovalent random immobilization is generally fulfilled by physical adsorption through hydrophobic, polar, or ionic interactions due to the simplicity and high amount of attached proteins. Covalent nonspecific random immobilization often utilizes the native functionalities of proteins, such as the thiol group of cysteine and the side-chain amino group of lysine. These functionalities are capable of reacting with surfaces modified by aldehydes, epoxides, maleimides, or N-hydroxysuccinimide (NHS) esters (Dai et al. 2020; Silva et al. 2021; Mann et al. 2020). Although the advantages of achieving protein immobilization without the requirements of extensive protein modification, these straightforward methods suffer from the loss of binding sites and protein activities, thus resulting in false-negative signals and heterogeneous data.

This review intends to provide an overview of the currently available techniques and the state of art of oriented protein immobilization strategies. Most such methodologies are developed through complementary affinity interactions occurring in almost all biological processes. For instance, polyhistidine and metal ions, avidin and biotin, antibodies and antigens, lectins and free saccharidic chains or glycosylated macromolecules, nucleic acids, and nucleic acid-binding proteins, hormones and their receptors, etc. (Redeker et al. 2013). Owning to the high specificity and affinity of these interactions, such techniques are advantageous to fabricate homogeneous surfaces with controllable protein orientation and higher protein activity.

3.1.1   Histidine Tag

As the most well-known affinity tag, His-tag was first fully realized by Hochuli in 1988 (Hochuli 1988). This small tag, usually comprising six sequential histidine residues, chelates transition metals including Cu(II), Co(II), Zn(II), or Ni(II). In this case, support is normally activated by nitrilotriacetic acid (NTA) or iminodiacetic acid. The such modification provides a chelating moiety for the metal ions when the support is treated with a solution of the relevant metal salt. The metal-activated support is subsequently used for immobilization of protein recombinantly bearing the tag through affinity interaction between the metals and the tag (Fig. 3.1a).

An example of applying His-tag in the immobilization of GPCR has been reported by Vogel group (Schmid et al. 1998) where the authors have successfully immobilized his-tagged detergent-solubilized serotonin-gated ion channel 5-HT$_3$R on nitrilotriacetic acid-modified quartz slides through metal chelate. By total internal reflection fluorescence, they observed demonstrable ligand-binding activity of the immobilized receptor without the need for further lipid reconstitution. Besides extensive attachment of 5-HT$_3$R, this approach has been broadly applied to the immobilization of several GPCRs, including the nicotinic acetylcholine receptor (nAChR) on glass surfaces (Rigler et al. 2004) and olfactory receptors on gold surfaces (Guo et al. 2015).

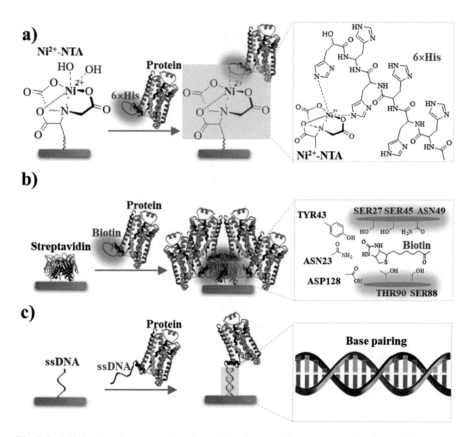

Fig. 3.1 Affinity-based non-covalent immobilization. **a** His-tagged protein immobilization; **b** biotin-streptavidin interaction-based immobilization; **c** DNA-mediated immobilization

This procedure is extremely simple and can be applied in a variety of immobilized GPCR-based analytical techniques, given that the receptor can be functionally solubilized and expressed with a His-tag. The advantages include a high-yield capacity, a high-salt loading of the lysates without losing binding affinities, and multiple regeneration cycles of the support when competitive ligands including imidazole and histidine or metal chelators like ethylenediaminetetraacetic acid (EDTA) are applied to the support. Besides the disadvantageous of stability, this approach appears to have high, nonspecific, but cooperative affinity to DNA and DNA-binding proteins (Banerjee et al. 2014) as well as some cellular proteins (Graslund et al. 2008), including many membrane (Kumazaki et al. 2014a, b) and eukaryotic proteins (Ohana et al. 2011).

3.1.2   Peptide Tags

The high specific affinity between epitope and antibody has offered an attractive alternative for protein immobilization. Although in principle any short stretch of amino acids can serve as epitope tags, most of them are constructed of short peptides (Table 3.1). Most widely recognized as FLAG, c-Myc, and HA tags, these epitope tags are typically eight to twelve amino acids in length with a small molecular weight of 0.6–6 kDa, and have very strong and specific affinity to their corresponding immunoaffinity resins. Such characteristics allow them to retain many of the advantages of the His-tag while providing superior purity and recovery of the fused target. Among these commonly used tags, FLAG-tag is a highly charged and soluble affinity tag composed of short octapeptide (Asp-Tyr-Lys-Asp-Asp-Asp-Asp-Lys) (Wang et al. 2001a, b). Owning to the high hydrophilicity, the FLAG tag usually locates on the fusion protein surface, therefore making the tag attractive for oriented immobilization. Taking the advantage of high specific affinity to several commercial anti-FLAG poly- and monoclonal antibodies, documented reports have utilized FLAG tag to immobilize proteins by their C terminus to solid supports coated with anti-FLAG antibodies (Wang et al. 2001a; Sydor et al. 2003). These immobilized proteins have proved to not only present higher activity than their randomly immobilized counterpart but also show longer storage stability. A similar strategy has been extended to the immobilization of detergent-soluble fluorescent β_2-AR through its amino-terminal FLAG tag on a surface layered with M1 antibody. This makes it possible to study the conformational dynamics of single immobilized receptors and to generate arrays of functional GPCRs for novel high-throughput screening strategies (Neumann et al. 2002).

The Myc-tag is an epitope of eleven amino acids (Table 3.1) derived from the c-Myc gene after the murin anti-c-Myc antibody was developed in 1985 (Evan et al. 1985). This tag can be genetically placed to either the N- or C-terminus of target proteins without the loss of its affinity toward the anti-Myc antibody. Wingren et al. have reported the proof-of-concept for a protein microarray design based on Myc-tagged single-chain antibody fragments (Wingren et al. 2005). They immobilized probes through engineered C terminal Myc-tags to fabricate arrayed monoclonal

Table 3.1 Basic information of the peptide affinity tags

Tag	Res.	Sequence	Size (kDa)	Affinity KD (M)	Loading NaCl (M)	Capacity (mg/ml)	Ref. or Manu
Poly-His	2–10	HHHHHH	0.84	$10^{-8/-9}$	1.0+	20–40	Marina et al.
FLAG	8	DYKDDDDK	1.01	$10^{-8/-9}$	<0.2	0.6	Marina et al.
Strep-tag II	8	WSHPQFEK	1.06	10^{-6}	<0.2	6.0	Marina et al.
HA	9	YPYDVPDYA	1.10	$10^{-9/-10}$	<0.2	≥0.01	Thermo
VSV-G	11	YTDIEMNRLGK	1.34	$10^{-7/-9}$	<0.2	≥0.83	Sigma
c-Myc	11	EQKLISEEDL	1.20	$10^{-7/-8}$	<0.2	≥0.05	Takara
V5	14	GKPIPNPLLGLDST	1.42	$10^{-7/-8}$	<0.2	0.7	ChromoTek
S tag	15	KETAAAKFERQHMDS	1.75	N/A	<0.2	8.0	Novagen
HAT	19	KDHLIHNVHKEFHAHAHNK	2.31	N/A	<0.2	N/A	Marina et al.
3 × FLAG	22	DYKDHDGDYKDHDIDYKDDDDK	2.73	$10^{-9/-10}$	<0.2	≥0.6	Sigma
CalBP	26	KRRWKKNFIAVSAANRFKKISSSGAL	2.96	10^{-9}	~1.0	1.5	Marina et al.
ChBD	51	TNPGVSAWQVNTAYTAGQLVTYNGKTYKCLQPHTSLAGWEPSNVPALWQLQ	5.59	10^{-6}	~1.0	2.0	Marina et al.

(continued)

Table 3.1 (continued)

Tag	Res.	Sequence	Size (kDa)	Affinity KD (M)	Loading NaCl (M)	Capacity (mg/ml)	Ref. or Manu
Im7	87	MELKNSISDYTEAEFVQLLKEIEKENVAATDDVLDVLLEHFVKITEHPDGT DLIYYPSDNRDDSPEGIVKEIKEWRAANGKPGFKQG	9.89	$10^{-14/-17}$	1.0+	15–20	Marina et al.
CBD	27–189	Domains	3.00–20.00	N/A	<0.2	1.5–40	Novagen
GST	211	Protein	26.00	$10^{-6/-7}$	<0.2	10	Marina et al.
Pr-A	252	Protein	28.33	$10^{-8/-10}$	<0.2	2.0–3.0	Marina et al.
MBP	396	Protein	40.00	10^{-6}	<0.2	6–10	Marina et al.

Res.: number of residues; Ref.: reference; Manu.: manufacture
Ref.: Marina N. Vassylyeva, Sergiy Klyuyev, Alexey D. Vassylyev, Hunter Wesson, Zhuo Zhang, Matthew B. Renfrow, Hengbin Wang, N. Patrick Higgins, Louise T. Chow, and Dmitry G. Vassylyev, Efficient, ultra-high-affinity chromatography in a one-step purification of complex proteins, PNAS, 2017, E5138–E5147

anti-tag antibodies. The functional antibody microarrays exhibit high specificity and sensitivity with a limit of detection in the pM to fM range (Wingren et al. 2005). Other epitope tags include the HA-tag (Tyr-Pro-Tyr-Asp-Val-Pro-Asp-Tyr-Ala) from influenza virus hemagglutinin (Rini et al. 1992) and the V5-tag (Gly-Lys-Pro-Ile-Pro-Asn-Pro-Leu-Leu-Gly-Leu-Asp-Ser-Thr) from the C-terminus of the P and V proteins of Simian virus 5 (Kolodziej et al. 2009). Although a successful application in protein immobilization, these epitope tags have not been applied in preparing immobilized membrane protein, especially GPCRs.

3.1.3 Biotin–(Strept) Avidin Interaction

Biotin–Streptavidin interaction is characterized by a very high affinity ($K_d = 10^{-15}$ M) (Grabowski and Sharp 1986). This strong interaction has been frequently utilized for the detection and purification of proteins (Yang et al. 2018; Kong et al. 2016), including integral membrane receptors. The small biotin tag can also serve as an anchor for the immobilization of protein onto solid surfaces (Fig. 3.1b). Commercial surfaces coated with streptavidin are readily available as chemical reagents to efficiently bind biotinylated proteins (Cho et al. 2020). The report that applies this method in the immobilization of enzymes has demonstrated a comparable activity to that of the wild-type enzyme in solution. The immobilized enzyme appears to be highly stable without loss of activity for over a week (Holland-Nell and Beck-Sickinger 2007; Yu et al. 2012).

With the aforementioned characteristics, the biotin-streptavidin interaction has become a relatively powerful alternative for GPCR immobilization. Successful application embraces oriented immobilization of rhodopsin (Hou et al. 2006), neurotensin receptor-1 (NTS-1) (Harding et al. 2006), peripheral cannabinoid receptor (CB2) (Krepkiy et al. 2006) and β_2-adrenergic receptor (β_2-AR) (Neumann et al. 2002). This method can implement the immobilization of soluble proteins with minimal loss due to the high-affinity coupling and does not require protein reconstitution into a lipid bilayer. However, the immobilization of GPCR by chemical biotinylation often results in an alteration of receptor structure and ligand binding parameters. This necessitates the development of a simpler procedure that allows efficient biotinylation of GPCRs with many hydrophobic domains.

3.1.4 DNA-Mediated Immobilization

The affinity of single-stranded DNA (ssDNA) to its complementary sequence is one of the strongest and most exquisitely specific interactions in nature (Kovacevic et al. 2018). This enormous specificity between complementary single-stranded nucleic acids is well exploited to establish powerful assays for analyzing DNA and RNA (Porto et al. 2020). In these assays, a labeled target is recognized and bound through predictable Watson–Crick base-pairing interactions by a surface-immobilized probe (Fig. 3.1c). Such principle has been successfully extended to protein immobilization where conjugation of proteins with artificial nucleic acids is achieved by modifying the proteins with a robust DNA tag and complementary ssDNA capture strands are covalently immobilized on the solid substrates to enable reversible anchoring of the protein conjugate (Leidner et al. 2019; Arrabito et al. 2013; Becker et al. 2005). For instance, the work by Francis, M. B. et al. has modified a glass surface with ssDNA and site-selectively attached the complementary DNA strand to the N-terminus of the aldolase enzyme (Palla et al. 2017). Their results have validated that the DNA strands serve as easily tunable and reversible chemical handles to hybridize the protein–DNA conjugates onto the surface; the immobilized enzyme retains its catalytic activity and the surfaces are reusable in subsequent cycles. To take full advantage of the DNA-directed strategy in protein immobilization, an interesting work by Niemeyer, C. M. et al. have pre-modified polymer films with ssDNA capture strands and microthermal-formed the resulting films into 3D structures followed by post-modification with complementary DNA–protein conjugates to realize complex biologically active surfaces within microfluidic devices (Schneider et al. 2017). They have found that microchannels presenting three different proteins on their inner curvilinear surface can be used for the selective capture of cells under flow conditions (Schneider et al. 2017).

The particular advantage of DNA-directed immobilization is the ease of sequence design and preparation in combination with the very high specificity of Watson–Crick base pairing. Despite tremendous advances in the preparation of DNA–protein conjugates, certain challenges still need to be overcome for site-specific coupling of the DNA to the protein and the feasibility of such a strategy in GPCR immobilization.

3.1.5 Glutathione-S-Transferase (GST) Tag

Other affinity strategies that have been utilized for protein immobilization include the interactions of GST to glutathione (Waugh 2005), antibodies to protein A or protein G (Han et al. 2020), maltose-binding protein to maltose (Seo et al. 2014), chitin-binding protein to chitin (Sanchez-Vallet et al. 2015). Successful as they work in protein purification, these methodologies are undesirable for protein immobilization in the final application. In the most widely-accepted protocol, proteins fused with the tag are often trapped on a gel column modified by the specific tag partner for the pursuit of purification and are then eluted with a solution containing the tag or its derivatives. The removal of the tag on the eluted protein is too hard to obstruct the following immobilization. Even if the tag is removed by size exclusion chromatography, such strategies are unfavorable in GPCR immobilization because the relatively weak nature of the binding requires a long time and intensive labor to implement the receptor immobilization. This may become a major disadvantage ascribed to the increasing possibility of receptor misfolding and the loss of activity.

3.2 Site-Specific Covalent Immobilization Through Bioorthogonal Chemistry

Despite great contribution to protein purification and immobilization, non-covalent immobilization by diverse affinity-based strategies is commonly disadvantageous attributed to two reasons: the low selectivity arising from relatively weak affinity between the tags and their partners often results in time-consuming and labor-intensive purification to pursue the removal of co-existing protein; the repeatability of the methods needs further improvement due to the uncontrollable density and amount of the immobilized protein. As an alternative, oriented covalent immobilizing assays have been successfully applied to address these issues owning to the precise control of immobilized protein density, the formation of homogenous and oriented protein layer on solid supports as well as the accessibility to the binding site after oriented immobilization. In light of these advantages, site-specific covalent protein immobilization has recently attracted substantial attention from scientists in diverse disciplines.

Several chemistries are capable of achieving an oriented covalent immobilization of proteins. In this context, bioorthogonal chemical reactions allow the regioselective attachment of small-molecule probes to protein and have become an exceedingly popular tool for site-specific covalent protein immobilization in both academic and industrial research. The characteristics of this strategy consist of mild reaction conditions, the utilization of water as a reaction medium, the stability in oxygen and water environment, and the high yield formation of a stable product under physiological conditions. Typically, the immobilization has been accomplished by two key steps (Fig. 3.2a): the site-specific modification of proteins with non-native chemical functionalities or tags; the attachment of the conjugated proteins to surfaces containing desired functional groups through bioorthogonal chemical reactions or enzyme-mediated transformations. A few reactions including Staudinger-ligations, Diels–Alder cycloadditions, oxime formation, and thiol-ene additions and copper or ring-strain catalyzed azide–alkyne cycloadditions, have been successfully developed to realize such bioorthogonal chemical protein immobilization (Chen et al. 2011).

3.2.1 Staudinger Ligation

To pursue selective covalent protein immobilization, diverse assays have been developed to incorporate biocompatible non-native functional groups into proteins. These assays consist of native chemical ligation (Cistrone et al. 2019), amber codon suppression mutagenesis (Zhu et al. 2014), metabolic engineering (Wang et al. 2017), and enzymatic attachment (Zhang et al. 2015). Few of these methods meet the requirements of GPCR immobilization when it comes to a high-yielding, robust, fast, and chemoselective method using low protein concentration and reacting conditions compatible with the physiological environment.

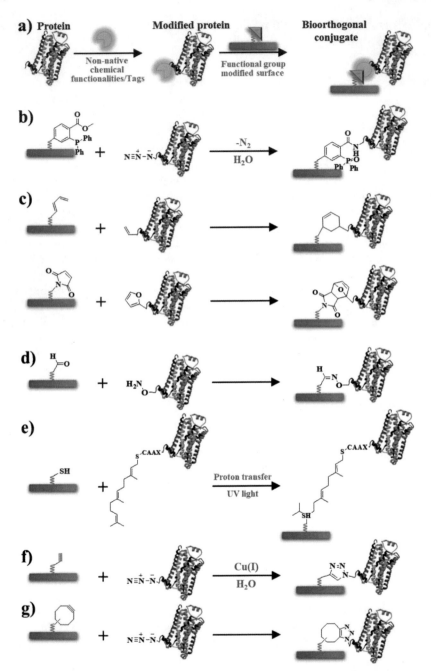

Fig. 3.2 Site-specific covalent protein immobilization. **a** Key steps for site-specific covalent protein immobilization; **b** staudinger-ligation; **c** diels–alder reactions; **d** oxime ligation; **e** thiol-ene additions; **f** copper(I)-catalyzed azide-alkyne cycloaddition; **g** ring-strain catalyzed azide–alkyne cycloaddition

The pioneering work by Staudinger, H has discovered a reaction that is able to produce an iminophosphorane by the reduction of an azide to an amine using phosphine or phosphite (Staudinger and Meyer 1919). This work has been enormously improved until Bertozzi and Saxon develop a triaryl phosphine reagent in which an electrophilic carbonyl group (usually a methyl ester) is introduced to trap the aza-ylide intermediate for the formation of a stable amide bond in aqueous media (Saxon and Bertozzi 2000) (Fig. 3.2b). This modified Staudinger reaction is widely termed as Staudinger ligation because it ligates two molecules together. Afterward, Staudinger ligation has been broadly applied in labeling biomolecules both in vitro and in vivo due to the exquisite bioorthogonal of azides and triarylphosphines, the high ligation yield, and the mild reaction conditions (Wen et al. 2018; Alamudi et al. 2018). Subsequent exploration of such strategy by Raines, R. T. et al. has demonstrated an effective means for the site-specific, covalent immobilization of a protein whereby an immobilization yield of >50% is obtained in <1 min, and immobilized proteins have >80% of their expected activity (Soellner et al. 2003). Further work by the same group has enabled Staudinger ligation to become a general method for site-specific protein immobilization (Kalia et al. 2007; Tam et al. 2007). Limitations of Staudinger ligation include the oxidation of phosphine reagents easily oxidize in air, the relatively slow kinetics of the reaction, and the adverse affection of attached diphenylphosphine oxide on conjugate solubility in water (Sletten and Bertozz 2009).

3.2.2 Diels–Alder Reaction

The Diels–Alder reaction often occurs between a dienophile and a diene to produce an unsaturated six-membered ring (Wang et al. 2016) (Fig. 3.2c). By this reaction, a gold-thiol self-assembled monolayer in the presence of quinones has been utilized to immobilize synthesized short peptides equipped with a peptidecyclopentadiene moiety (Dillmore et al. 2004). This method is advantageous due to the elimination of slide-blocking procedures and the provision of a regular, homogeneous environment for immobilized peptide ligands. Despite the potential to become well-suitable quantitative assays, this method is challenged by the synthesis of peptides conjugated to an unnatural cyclopentadiene moiety.

Site-specific immobilization of entire functional protein using Diels–Alder ligation has also been successfully implemented by employing streptavidin as a representative biologically relevant example (de Araujo et al. 2005). In this case, diene-functionalized streptavidin (positive control) is prepared with a cyclopentadiene derivative and spotted on a maleimide-modified microarray slide using non-modified streptavidin as a negative control. Subsequent treatment of the slides is performed by fluorophore-labeled Cy5-biotin. Further detection by fluorescence scanning exhibits intense fluorescence signals of immobilized diene-modified streptavidin as well as unobservable signals of the negative control. Owning to the high biocompatibility, selectivity, and efficiency in an aqueous solution, this method has allowed the successful immobilization of diverse proteins on glass slides. However, this method is partly limited by the reversibility of the reaction, hydrolysis susceptibility of the dienophiles, and slow reaction kinetics (Fisher et al. 2017), which has not yet been applied to full-length biologically active GPCRs.

3.2.3 Oxime Ligation

The reaction between ketone/aldehyde and amine nucleophiles to form an oxime linkage has proved to be enhanced by α-effect like alkoxyamines and hydrazines (Fig. 3.2d). In some cases, oxime ligation is limited by the utilization of carbonyl compounds due to the slow rate constants, instability of the hydrazone and oxime bonds as well as the competition with endogenous aldehydes and ketones of biomolecules inside cells (Rashidian et al. 2013). To address the issues of pH-depending reaction rate, aniline or m-phenylenediamine is introduced to catalyze the reaction under neutral conditions (Dirksen et al. 2006).

Despite the aforementioned limitations, oxime ligation has been elegantly explored for modification of biologics, as the oxyamine and ketone/aldehyde functionalities are biorthogonal and their conjugation proceeds site specifically with fast kinetics and high yield under physiological conditions (Tang et al. 2015). These advantages have also made oxime ligation successful application in the surface immobilization of proteins (Uth et al. 2014).

3.2.4 Photochemical Thiol-Ene Reaction

The thiol–ene reaction is capable to introduce a thiol to an ene group by a free radical mechanism (Lim et al. 2020). This reaction often proceeds at UV wavelengths of 365–405 nm to form a thioether bond with good stability. The advantages of this reaction include the achievement of quantitative yields, the rapid reaction rates either in bulk or in environmentally benign solvents, the unnecessary cleaning, the insensitivity to ambient oxygen or water, and the yield of a single regioselective product (Hoyle and Bowman 2010). These exceptional versatilities make thiol–ene reaction amenable to applications in the optical, biomedical, and bioorganic modification fields. Owning to the specificity of olefins and robustness in aqueous buffer, the coupling of thiols and olefins induced by photochemistry has been extended to surface patterning and immobilization of biomolecules on solid supports (Gobel et al. 2014). By this reaction, biotin with an olefin derivative has been attached to a thiol-functionalized surface. Subsequent treatment by Cy5-labeled streptavidin produces a fluorescent protein-patterned surface. Such a surface proves to be applicable in anchoring biotinylated alkaline phosphatase through biotin-streptavidin binding. Demonstrable enzymatic activity of the immobilized enzyme is observed by fluorescence assays using the substrates like Vector Red. Following experiments using the Ras binding domain of cRaf1 and GppNHp as model pairs have also demonstrated the functionality of the surface-immobilized proteins.

Further research have applied the photochemical thiol-ene reaction in direct and covalent immobilization of proteins. This generally requires the incorporation of an olefin group into the protein. Taking inspiration from three olefins in the farnesyl group, the pioneering work has introduced olefins into H-Ras, N-Ras, and K-Ras bear the CAAX box by farnesyl transferase and has realized the direct immobilization of the three proteins on thiol-functionalized chips upon exposure to UV light at a wavelength of 365 nm for 10 min (Weinrich et al. 2010) (Fig. 3.2e). Using a fluorescent Cy3-labeled antibody against Ras isoforms, the authors observed demonstrable Ras-positive microstructures. This proves the successful immobilization and correct folding of the Ras proteins. Applying the method in lysates of *E. coli* cells, the author reached thiol-ene photo-immobilization of target proteins without purification.

These investigations have demonstrated the general applicability of photochemical thiol-ene reactions. It has the potential to provide a strategy for enhancing the immobilization of other proteins like GPCR to fabricate immobilized GPCR-based bioship, microarray, and stationary phase.

3.2.5   Copper(I)-Catalyzed Azide-Alkyne Cycloaddition

As an archetypal example of click chemistry, the copper-catalyzed Huisgen 1,3-dipolar cycloaddition of azides and alkynes has been known to form 1,2,3-triazole (Stenehjem et al. 2013; Astruc et al. 2012; Mamidyala et al. 2010). Apart from the direct use of several Cu(I) sources, many protocols prepare the catalyst in situ by reduction of Cu(II) salts (Fig. 3.2f). Such mandatory copper catalyst often causes proteins to precipitate. This, to a large extent, reduces the broad application of the reaction for protein immobilization. Such an issue has been partly addressed until Cho et al. has devised a bioorthogonal reaction between sulfonyl azides and terminal alkynes for amide synthesis in 2005 (Cho et al. 2005). Intrigued by the report, Fokin's group has studied reactions of sulfonyl azides with copper(I) acetylides like terminal alkynes and sulfonyl azides, and have found that copper(I) ions efficiently catalyze the direct formation of pure N-acylsulfonamide products in good yield without complicated purification (Cassidy et al. 2006). These results have demonstrated that copper(I)-catalyzed azide-alkyne cycloaddition in terms of sulfonamide reaction is high chemo-selective and robust under mild conditions and is readily amenable to the solid-phase procedure.

The application of sulfonamide reaction in protein immobilization has been first performed by Govindaraju et al. in 2008 (Govindaraju et al. 2008). Initially, they introduce an alkyne moiety to modify small biomolecules such as biotin, R-D-mannose, and phosphopeptides. These modified biomolecules are then ligated with a benzyloxycarbonyl-protected tauryl sulfonyl azide in an aqueous solution in the presence of Cu(I)-catalyst. Owning to a 90–95% conversion within 2–4 h, the authors declare that sulfonamide reaction has potential in surface immobilization. To prove their declaration, they introduce an alkyne moiety at the C-terminus of a fluorescent mCherry-Ypt7 protein by expressed protein ligation. Spotting the alkyne-modified and the nonmodified protein onto sulfonyl azide-functionalized slides, the authors observe clear fluorescence signals of mCherry-Ypt7 embodying an alkyne group. By similar protocol, the authors also apply sulfonamide reaction in surface immobilization of the Ras-binding domain of cRaf1. Taking together, the authors confirm that alkyne-modified biomolecules can be immobilized site- and chemo-selectively on sulfonyl azide surface under very mild conditions by sulfonamide reaction.

All the aforementioned bioorthogonal ligation methods are covalent, site- and chemo-selective to fulfill protein immobilization without losing protein three-dimensional structure and activity. The choice of an exact method for a particular case depends on the properties of the target protein. The photochemical thiol-ene reaction should be avoided if the protein to be immobilized is not stable under UV radiation. Similarly, the sulphonamide reaction is not the right choice if the protein to be immobilized is sensitive to copper, as this immobilization is catalyzed by the metal. For unstable proteins like GPCR, methods characterized by short reaction times are preferable to be applied to minimize the loss of protein three-dimensional structure and activity. In this case, the photochemical thiol-ene reaction (30 min), the Staudinger ligation (4 h), or the "click sulfonamide" reaction (4 h) appear to be more suitable than the Diels–Alder cycloaddition (8 h).

3.3   Biologically Mediated Site-Specific Immobilization

The advances in biotechnology allow several chemical and biological techniques to introduce bioorthogonal functional groups into proteins. Among them, the most widely applied and easiest way appears to be the coupling of a short bifunctional cross-linker to the unfunctional endogenous amino acids of the protein. In most cases, this is performed with the utilization of lysines or cysteines. Like the above-mentioned covalent immobilization, this method often suffers from the multiplicity of the endogenous functional groups, thereby limiting the possibility to introduce the bifunctional linker to the protein in a site-specific way. This often leads to undesired modifications of the protein through the active or specific binding sites. These challenges are partly overcomed by chemoenzymatic methods that allow site-specific modification of the protein with special labels (Li et al. 2021; Hofmann et al. 2020; Schumacher et al. 2018). As an increasingly active research field, such methods prove to be highly selective for site-specific protein immobilization owning to the exquisite selectivity of enzymes. By the role of the enzyme in the immobilization procedure, successful cases can be classified into enzyme-catalyzed immobilization, self-labeling tags, and native chemical ligation.

3.3.1 Enzyme-Catalyzed Immobilization

The enzyme-catalyzed reaction has been extensively reported to achieve great reaction yield, faster reaction rate, high selectivity, and good compatibility with complex biological systems. These advantages enable enzymes, in particular enzymes involved in post-translational modification, a preferable choice to introduce specific functionalities into proteins for subsequent immobilization. In this context, the pioneering work by Chen et al. (2005) has reported that the Escherichia coli enzyme biotin ligase (BirA) catalyzes the sequence-specific biotinylation of a 15 amino acid peptide at its lysine side chain. This reaction can be accomplished in 20 min without substantial influence on the specificity of small changes in the biotin structure (Slavoff et al. 2008). Taking epidermal growth factor receptor (EGFR) as an example, they have confirmed that proteins tagged with the peptide are readily to be selectively biotinylated and coupled to streptavidin-modified support under BirA catalysis.

Sortase A (SrtA) is an enzyme that catalyzes the covalent capture of proteins containing a recognition motif of LPXTG (X can be any amino acid) to cell wall peptidoglycan (Cao et al. 2018). Applications of this specific reaction have been extensively explored in covalent protein immobilization on solid surfaces (Jiang et al. 2012). Recently, Krause, E. et al. have utilized the specificity of SrtA transpeptidation reaction to prepare affinity matrices in which a protein bait is covalently linked to the matrix via a short C-terminal linker region (Kuropka et al. 2015). They demonstrate that this site-directed immobilization enables the bait to remain functionally accessible to protein interaction (Kuropka et al. 2015). Their pull-down experiments confirm the suitability of the method for identification of protein interactions in affinity purification–mass spectrometry experiments (Kuropka et al. 2015). Other enzymes that have been widely applied for protein ligation are providing powerful strategies for covalent protein immobilization, including protein farnesyltransferase (Gauchet et al. 2006), transglutaminase (Giannetto et al. 2014), and lipoic acid ligase (Plaks et al. 2015). Even though the lack of proof in GPCR immobilization, these reports have paved the way to propose a universally accepted protocol for enzyme-catalyzed immobilization. It needs three steps: incorporating a specific tag into the protein of interest, coupling the substrate of the enzyme to the solid support, and catalyzing the immobilization by the enzyme.

3.3.2 Self-Labelling Active Enzyme

In addition to catalyzing immobilization, enzymes have also been performed as self-labeling tags to attach proteins onto a solid surface. Such tags dominantly originated from enzymes that covalently react with a substrate to form a stable product. The enzyme itself is often fused directly to the protein of interest, while its substrate is linked to a solid surface. This makes an easier system than enzyme-catalyzed immobilization to pursue the attachment of the target protein.

Human DNA repair protein O^6-alkylguanine-DNA alkyltransferase (hAGT) is a 20 kDa protein that transfers the alkyl group from its substrate, O^6-alkylguanine-DNA, to an internal reactive cysteine residue of hAGT (Kindermann et al. 2003). Johnsson et al. first reported a general method for the immobilization of proteins that are genetically linked to a mutant of the hAGT (Keppler et al. 2003). This enables fusion proteins to selectively react with surfaces displaying O^6-benzylguanine (BG) derivatives, producing specific, covalent, and quasi-irreversible immobilization. Owning to the high specificity for hAGT but promiscuous with respect to the substrate, this strategy motivates continuous attempts to immobilize hAGT-tagged proteins onto a solid surface for diverse purposes. Recent successful cases consist of the fabrication of self-assembled monolayers for protein immobilization of green fluorescence protein (Ul-Haq et al. 2013), Sonic hedgehog (Kwok et al. 2011), cytokines (Casalini et al. 2015) and antibodies (Hussain et al. 2019).

Our recent work introduces hAGT into the immobilization of GPCRs taking beta2-adrenergic receptors (β_2-AR) as an example due to its well-acknowledged crystal structure and functions (Wang et al. 2019). We cloned hAGT as a tag on the C-terminal of β_2-AR primarily to express the fusion receptor in *E. coli*. To achieve the immobilization of β_2-AR, we applied the cell lysate directly to a BG derivative-modified PEGA (O^6-benzylguanine-modified polyethylene glycol polyacrylamide) resin. Characterization of the immobilized receptor displayed even distribution and remarkable ligand-binding activity of β_2-AR on the resin surface. Applying this immobilized β_2-AR as a stationary phase, we established an affinity-selected-mass-spectrometric system.

Another self-labeling enzyme, known as HaloTag, is a mutant bacterial haloalkane dehalogenase enzyme from Rhodococcus rhodochrou. This enzyme forms an alkyl-enzyme intermediate during the nucleophilic displacement of a terminal chloroalkane or a bromoalkane with Asp106 (England et al. 2015). In the wild-type haloalkane dehalogenases, His272 functions as a general base to catalyze the hydrolysis, thereby releasing the enzyme and hydrocarbon as alcohol. Such hydrolysis is blocked by the replacement of His272 with Phe272, thus trapping the reaction intermediate as a stable covalent adduct (Los et al. 2008). In the HaloTag, the modified enzyme contains a critical mutation in the catalytic site that traps the reaction intermediate as a stable covalent adduct. The formation of this covalent bond proves to be essentially irreversible with high specificity and rapid speed under physiological conditions. This makes HaloTag a versatile platform for biomedical applications including protein immobilization (Fu et al. 2021).

Our group has utilized haloalkane dehalogenase (Halo) as an immobilization tag fused to the β_2-AR, angiotensin II type 1, and angiotensin II type 2 receptors (Zeng et al. 2018). We express the three Halo-tagged GPCRs in Escherichia coli at a suitable yield and immobilize them onto macroporous silica gel coated with chloroalkane using the cell lysate of the three Halo-tagged GPCRs. Morphological characterization demonstrates a homogeneous monolayer of immobilized Halo-tagged GPCRs on the silica gel surface. Specific bound phospholipids including PG C18:1/C18:1 prove to surround the immobilized receptors. Probably, for this reason, we observe a radio-ligand binding ability and ligand-induced conformational changes in the immo-bilized GPCRs. This method is a one-step approach for the specific immobilization of GPCRs from cell lysates and validates that immobilized receptors retain canonical ligand-binding capacity. It circumvents labor-intensive purification procedures and minimizes loss of activity.

Our recent work introduces the kinase domain of epidermal growth factor receptor (EGFR), the amino acids between 694 to 1022, as a fusion tag to the endothelin receptor A (ETA) (Zhao et al. 2020). The kinase domain, namely EGFR-tag, can irreversibly interact with EGFR covalent inhibitor (ibrutinib) in a rapid way, owning to the formation of a covalent bond between the electrophilic Michael acceptor of the inhibitor and the conserved cysteine (Cys797) in EGFR. We expressed the EGFR-tagged ETA in *E. coli* and realized the GPCR immobilization by incubating the *E. coli* cell lysate with ibrutinib modified silica gel. The feasibility of the immobi-lized ETA was verified by a series of chromatographic methods for receptor-drug interaction analysis and lead identification from Siwu decoction. The EGFR-tag system has the advantages as follows: (i) the high specificity and covalent nature between ibrutinib and EGFR-tag greatly minimized the non-specific interactions with unwanted substances in the coexisting system, making the immobilization process in a robust, fast and mild fashion. (ii) The immobilized receptor has high specificity in recognizing its ligands and identifying lead compounds from complex system.

The merit of the strategies based on the self-labeling enzyme lay behind the simplification of the immobilization procedure and reaction system. However, they incorporate large tags (~20 kDa for hAGT, 33 kDa for the HaloTag, and 37 kDa for the EGFR tag) into the target protein, which may affect the functions of the protein. For this reason, self-labeling enzyme related methods are not suitable for the immobilization of small proteins, thus hampering their broad application.

3.3.3 Native Chemical Ligation

Expressed protein ligation (EPL), derived from native chemical ligation, is described as a broadly applicable method for protein engineering. This method allows recom-binant and synthetic polypeptides to be chemoselectively and regioselectively joined together by protein splicing. In the process, self-catalyzed intramolecular rearrange-ments of a precursor protein bring about the removal of an internal sequence (intein), and the ligation of the lateral sequences (N- and C-exteins) (Muir et al. 1998). In EPL, the protein of interest is recombinantly expressed as a fusion protein with an intein mutant that is defective in completion of the splicing reaction but can form thioester intermediate. In standard practice, the target protein is cleaved from the intein-chitin-binding domain after the removal of undesired proteins by DTT or 2-mercaptoethanol while the chimera remains on a chitin column. By using a thiol-containing peptide in the presence of a bioorthogonal group, the splicing process is capable of producing a stable peptide bond through spontaneous rearrangement, which results in a protein terminally modified with the bioorthogonal group (Muir 2003).

In previous reports, EPL has been adopted extensively in the field of biotechnology and chemical biology, for example, the specific modification of a protein with diverse molecules (Vila-Perello 2013; Flavell and Muir 2009; Bang et al. 2006; Lovrinovic et al. 2003). Recently, Chattopadhaya et al. have expanded the application of EPL strategies into the realm of microarrays (Chattopadhaya et al. 2009). Their approach entails the generation of proteins that are site-specifically biotinylated at the C-terminus via EPL reaction between the thioester moiety at the C-terminus of target proteins and cysteine-biotin. The proteins are then site-specifically immobilized on avidin-coated slides (Chattopadhaya et al. 2009). Another work by Lin, C. C. et al. has utilized EPL approaches to modify and immobilize sialyltransferase (PmST1) on PEGylated magnetic nanoparticles (MNPs) (Yu et al. 2012). Their results demonstrate the stronger activity of the immobilized PmST1 than the native free enzyme (Yu et al. 2012). Their subsequent work has attached a membrane-bound sialyltransferase via a biotin-labeled cysteine at the C-terminus to streptavidin functionalized MNPs (Yu et al. 2012). They state that the method provides an approximately twofold increase in activity compared to other immobilization methods on MNPs (Yu et al. 2012).

Despite the achievement of protein immobilization under mild ligation conditions, the application of EPL strategies remains in view of several techniques. First, it appears to be restricted to the utilization of N- or C-terminus of the protein. This is not possible when the termini are unavailable. Second, the size of the incorporated modifications is limited to 50 amino acids from the termini of the protein. Even though this can be addressed by combinational use of other ligation steps, it substantially increases the complexity of the synthesis (McGinty et al. 2008). The third principal limitation comes from the need for cysteine at the ligation junction. This is partly overcomed by the advances in the use of auxiliaries or desulfurization reactions that allow the ligation to occur at phenylalanine, alanine, and glycine (Wan and Danishefsky 2007). Finally, the co-existing reactions of the thioester make EPL unlikely to be re-engineered to proceed in vivo. Collectively, in a positive view, ongoing advances in the techniques and concepts of EPL promise to extend this strategy to immobilize complex proteins like GPCRs.

Each of the abovementioned methods has advantages and disadvantages that need to be taken into account. We envisage that the combination of EPL with photochemical thiol-ene reaction has the potential to serve as a powerful platform for site-specific covalent immobilization of GPCRs. Such a combination is advantageous because it allows immobilization of GPCRs directly from expression lysates without additional manipulation, thereby minimizing the loss of receptor activity and facilitating the efficient generation of immobilized protein-related approaches, including protein microarrays, biochips, and stationary phases. Owning to the small size of the incorporated molecule, this combinational immobilization strategy is promising to produce little interference in subsequent application of the immobilized receptor.

References

Alamudi SH et al (2018) A palette of background-free tame fluorescent probes for intracellular multi-color labelling in live cells. Chem Sci 9(8):2376–2383. https://doi.org/10.1039/c7sc04716a

Arrabito G et al (2013) Biochips for cell biology by combined dip-pen nanolithography and DNA-directed protein immobilization. Small 9(24):4243–4249. https://doi.org/10.1002/smll.201 300941

Astruc D et al (2012) Click dendrimers and triazole-related aspects: catalysts, mechanism, synthesis, and functions. A bridge between dendritic architectures and nanomaterials. Acc Chem Res 45(4):630–640. https://doi.org/10.1021/ar200235m

Banerjee R et al (2014) Optimization of recombinant mycobacterium tuberculosis RNA polymerase expression and purification. Tuberculosis 94(4):397–404. https://doi.org/10.1016/j.tube. 2014.03.008

Bang D et al (2006) Kinetically controlled ligation for the convergent chemical synthesis of proteins. Angew Chem Int Ed Engl 45(24):3985–3988. https://doi.org/10.1002/anie.200600702

Becker CF et al (2005) Direct readout of protein-protein interactions by mass spectrometry from protein-DNA microarrays. Angew Chem Int Ed Engl 44(46):7635–7639. https://doi.org/10.1002/ anie.200502908

Cao T et al (2018) Selective enrichment and quantification of N-terminal glycine peptides via sortase A mediated ligation. Anal Chem 90(24):14303–14308. https://doi.org/10.1021/acs.ana lchem.8b03562

Casalini S et al (2015) Multiscale sensing of antibody-antigen interactions by organic transistors and single-molecule force spectroscopy. ACS Nano 9(5):5051–5062. https://doi.org/10.1021/acs nano.5b00136

Cassidy MP et al (2006) Practical synthesis of amides from in situ generated copper(I) acetylides and sulfonyl azides. Angew Chem Int Ed Engl 45(19):3154–3157. https://doi.org/10.1002/anie. 200503805

Chattopadhaya S et al (2009) Use of intein-mediated protein ligation strategies for the fabrication of functional protein arrays. Methods Enzymol 462:195–223. https://doi.org/10.1016/S0076-687 9(09)62010-3

Chen I et al (2005) Site-specific labeling of cell surface proteins with biophysical probes using biotin ligase. Nat Methods 2(2):99–104. https://doi.org/10.1038/NMETH735

Chen YX et al (2011) Bioorthogonal chemistry for site-specific labeling and surface immobilization of proteins. Acc Chem Res 44(9):762–773. https://doi.org/10.1021/ar200046h

Cho SH et al (2005) Copper-catalyzed hydrative amide synthesis with terminal alkyne, sulfonyl azide, and water. J Am Chem Soc 127(46):16046–16047. https://doi.org/10.1021/ja056399e

Cho KF et al (2020) Proximity labeling in mammalian cells with TurboID and split-TurboID. Nat Protoc 15(12):3971–3999. https://doi.org/10.1021/10.1038/s41596-020-0399-0

Cistrone PA et al (2019) Native chemical ligation of peptides and proteins. Curr Protoc Chem Biol 11(1):e61. https://doi.org/10.1002/cpch.61

Dai W et al (2020) Structure-based design of antiviral drug candidates targeting the SARS-CoV-2 main protease. Science 368(6497):1331–1335. https://doi.org/10.1126/science.abb4489

de Araujo AD et al (2005) Diels-Alder ligation and surface immobilization of proteins. Angew Chem Int Ed Engl 45(2):296–301. https://doi.org/10.1002/anie.200502266

Dillmore WS et al (2004) A photochemical method for patterning the immobilization of ligands and cells to self-assembled monolayers. Langmuir 20(17):7223–7231. https://doi.org/10.1021/ la049826v

Dirksen A et al (2006) Nucleophilic catalysis of hydrazone formation and transimination: implications for dynamic covalent chemistry. J Am Chem Soc 128(49):15602–15603. https://doi.org/ 10.1021/ja067189k

England CG et al (2015) HaloTag technology: a versatile platform for biomedical applications. Bioconjug Chem 26(6):975–986. https://doi.org/10.1021/acs.bioconjchem.5b00191

Evan GI et al (1985) Isolation of monoclonal antibodies specific for human c-Myc proto-oncogene product. Mol Cell Biol 5(12):3610–3616. https://doi.org/10.1128/MCB.5.12.3610

Fisher SA et al (2017) Designing peptide and protein modified hydrogels: selecting the optimal conjugation strategy. J Am Chem Soc 139(22):7416–7427. https://doi.org/10.1021/jacs.7b00513

Flavell RR, Muir TW (2009) Expressed protein ligation (EPL) in the study of signal transduction, ion conduction, and chromatin biology. Acc Chem Res 42(1):107–116. https://doi.org/10.1021/ar800129c

Fu X et al (2021) Halo-tagged protein immobilization: effect of halide linkers on peak profile and drug-protein interaction. J Chromatogr A 1640:461946. https://doi.org/10.1016/j.chroma.2021.461946

Gauchet C et al (2006) Regio- and chemoselective covalent immobilization of proteins through unnatural amino acids. J Am Chem Soc 128(29):9274–9275. https://doi.org/10.1021/ja061131o

Giannetto M et al (2014) An amperometric immunosensor for diagnosis of celiac disease based on covalent immobilization of open conformation tissue transglutaminase for determination of anti-tTG antibodies in human serum. Biosens Bioelectron 62:325–330. https://doi.org/10.1016/j.bios.2014.07.006

Gobel R et al (2014) Modular thiol-ene chemistry approach towards mesoporous silica monoliths with organically modified pore walls. Chem-Eur J 20(52):17579–17589. https://doi.org/10.1002/chem.201403982

Govindaraju T et al (2008) Surface immobilization of biomolecules by click sulfonamide reaction. Chem Commun 32:3723–3725. https://doi.org/10.1039/b806764c

Grabowski PJ, Sharp PA (1986) Affinity chromatography of splicing complexes: U2, U5, and U4 + U6 small nuclear ribonucleoprotein particles in the spliceosome. Science 233(4770):1294–1299. https://doi.org/10.1126/science.3638792

Graslund S et al (2008) Protein production and purification. Nat Methods 5(2):135–146. https://doi.org/10.1038/nmeth.f.202

Guo Z et al (2015) A novel platform based on immobilized histidine tagged olfactory receptors, for the amperometric detection of an odorant molecule characteristic of boar taint. Food Chem 184:1–6. https://doi.org/10.1016/j.foodchem.2015.03.066

Han X et al (2020) Hapten-branched polyethylenimine as a new antigen affinity ligand to purify antibodies with high efficiency and specificity. ACS Appl Mater Interfaces 12(52):58191–58200. https://doi.org/10.1021/acsami.0c15586

Harding PJ et al (2006) Direct analysis of a GPCR-agonist interaction by surface plasmon resonance. Eur Biophys J 35(8):709–712. https://doi.org/10.1007/s00249-006-0070-x

Hochuli E (1988) Large-scale chromatography of recombinant proteins. J Chromatogr 444:293–302. https://doi.org/10.1016/S0021-9673(01)94032-4

Hofmann R et al (2020) Lysine acylation using conjugating enzymes for site-specific modification and ubiquitination of recombinant proteins. Nat Chem 12(11):1008–1015. https://doi.org/10.1038/s41557-020-0528-y

Holland-Nell K, Beck-Sickinger AG (2007) Specifically immobilised Aldo/Keto reductase AKR1A1 shows a dramatic increase in activity relative to the randomly immobilised enzyme. ChemBioChem 8(9):1071–1076. https://doi.org/10.1002/cbic.200700056

Hou Y et al (2006) Immobilization of rhodopsin on a self-assembled multilayer and its specific detection by electrochemical impedance spectroscopy. Biosens Bioelectron 21(7):1393–1402. https://doi.org/10.1016/j.bios.2005.06.002

Hoyle CE, Bowman CN (2010) Thiol-ene click chemistry. Angew Chem Int Ed Engl 49(9):1540–1573. https://doi.org/10.1002/anie.200903924

Hussain AF et al (2019) One-step site-specific antibody fragment auto-conjugation using SNAP-tag technology. Nat Protoc 14(11):3101–3125. https://doi.org/10.1038/s41596-019-0214-y

Jiang R et al (2012) End-point immobilization of recombinant thrombomodulin via sortase-mediated ligation. Bioconjug Chem 23(3):643–649. https://doi.org/10.1021/bc200661w

Kalia J et al (2007) General method for site-specific protein immobilization by Staudinger ligation. Bioconjug Chem 18(4):1064–1069. https://doi.org/10.1021/bc0603034

Keppler et al (2003) A general method for the covalent labeling of fusion proteins with small molecules in vivo. Nat Biotechnol 21:86–89. https://doi.org/10.1038/nbt765

Kindermann M et al (2003) Covalent and selective immobilization of fusion proteins. J Am Chem Soc 125(26):7810–7811. https://doi.org/10.1021/ja034145s

Kolodziej KE et al (2009) Optimal use of tandem biotin and V5 tags in ChIP assays. BMC Mol Biol 10:6. https://doi.org/10.1186/1471-2199-10-6

Kong J et al (2016) Quantifying nanomolar protein concentrations using designed DNA carriers and solid-state nanopores. Nano Lett 16(6):3557–3562. https://doi.org/10.1021/acs.nanolett.6b00627

Kovacevic KD et al (2018) Pharmacokinetics, pharmacodynamics and safety of aptamers. Adv Drug Deliv Rev 134:36–50. https://doi.org/10.1016/j.addr.2018.10.008

Krepkiy D et al (2006) Bacterial expression of functional, biotinylated peripheral cannabinoid receptor CB2. Protein Expr Purif 49(1):60–70. https://doi.org/10.1016/j.pep.2006.03.002

Kumazaki K et al (2014a) Crystallization and preliminary X-ray diffraction analysis of YidC, a membrane-protein chaperone and insertase from bacillus halodurans. Acta Crystallogr F Struct Biol Commun 70(Pt8):1056–1060. https://doi.org/10.1107/S2053230X14012540

Kumazaki K et al (2014b) Structural basis of Sec-independent membrane protein insertion by YidC. Nature 509(7501):516–520. https://doi.org/10.1038/nature13167

Kuropka B et al (2015) Sortase A mediated site-specific immobilization for identification of protein interactions in affinity purification-mass spectrometry experiments. Proteomics 15(7):1230–1234. https://doi.org/10.1002/pmic.201400395

Kwok CW et al (2011) Selective immobilization of sonic hedgehog on benzylguanine terminated patterned self-assembled monolayers. Biomaterials 32(28):6719–6728. https://doi.org/10.1016/j.biomaterials.2011.05.069

Leidner A et al (2019) Oriented immobilization of a delicate glucose-sensing protein on silica nanoparticles. Biomaterials 190:76–85. https://doi.org/10.1016/j.biomaterials.2018.10.035

Li C et al (2021) Site-selective chemoenzymatic modification on the core fucose of an antibody enhances its Fcγ receptor affinity and ADCC activity. J Am Chem Soc 143(20):7828–7838. https://doi.org/10.1021/jacs.1c03174

Lim KS et al (2020) Fundamentals and applications of photo-cross-linking in bioprinting. Chem Rev 120(19):10662–10694. https://doi.org/10.1021/acs.chemrev.9b00812

Los GV et al (2008) HaloTag: a novel protein labeling technology for cell imaging and protein analysis. ACS Chem Biol 3(6):373–382. https://doi.org/10.1021/cb800025k

Lovrinovic M et al (2003) Synthesis of protein-nucleic acid conjugates by expressed protein ligation. Chem Commun 7:822–823. https://doi.org/10.1039/b212294d

Mamidyala SK et al (2010) In situ click chemistry: probing the binding landscapes of biological molecules. Chem Soc Rev 39(4):1252–1261. https://doi.org/10.1039/b901969n

Mann FA et al (2020) Quantum defects as a toolbox for the covalent functionalization of carbon nanotubes with peptides and proteins. Angew Chem Int Ed Engl 59(40):17732–17738. https://doi.org/10.1002/anie.202003825

McGinty RK et al (2008) Chemically ubiquitylated histone H2B stimulates hDot1L-mediated intranucleosomal methylation. Nature 453(7196):812–816. https://doi.org/10.1038/nature06906

Muir TW et al (1998) Expressed protein ligation: a general method for protein engineering. Proc Natl Acad Sci U S A 95(12):6705–6710. https://doi.org/10.1073/pnas.95.12.6705

Muir TW (2003) Semisynthesis of proteins by expressed potein ligation. Annu Rev Biochem 72:249–289. https://doi.org/10.1146/annurev.biochem.72.121801.161900

Neumann L et al (2002) Functional immobilization of a ligand-activated G-protein-coupled receptor. ChemBioChem 3(10):993–998. https://doi.org/10.1002/1439-7633(20021004)3:10%3c993::AID-CBIC993%3e3.0.CO;2-Y

Ohana RF et al (2011) HaloTag-based purification of functional human kinases from mammalian cells. Protein Expr Purif 76(2):154–164. https://doi.org/10.1016/j.pep.2010.11.014

Palla KS et al (2017) Site-selective oxidative coupling reactions for the attachment of enzymes to glass surfaces through DNA-directed immobilization. J Am Chem Soc 139(5):1967–1974. https://doi.org/10.1021/jacs.6b11716

Plaks JG et al (2015) Multisite clickable modification of proteins using lipoic acid ligase. Bioconjug Chem 26(6):1104–1112. https://doi.org/10.1021/acs.bioconjchem.5b00161

Porto EM et al (2020) Base editing: advances and therapeutic opportunities. Nat Rev Drug Discov 19(12):839–859. https://doi.org/10.1038/s41573-020-0084-6

Rashidian M et al (2013) A highly efficient catalyst for oxime ligation and hydrazone-oxime exchange suitable for bioconjugation. Bioconjug Chem 24(3):333–342. https://doi.org/10.1021/bc3004167

Redeker ES et al (2013) Protein engineering for directed immobilization. Bioconjug Chem 24(11):1761–1777. https://doi.org/10.1021/bc4002823

Rigler P et al (2004) Controlled immobilization of membrane proteins to surfaces for Fourier transform infrared investigations. Langmuir 20(19):7901–7903. https://doi.org/10.1021/la049002d

Rini JM et al (1992) Structural evidence for induced fit as a mechanism for antibody-antigen recognition. Science 255(5047):959–965. https://doi.org/10.1126/science.1546293

Sanchez-Vallet A et al (2015) The battle for chitin recognition in plant-microbe interactions. FEMS Microbiol Rev 39(2):171–183. https://doi.org/10.1093/femsre/fuu003

Saxon E, Bertozzi CR (2000) Cell surface engineering by a modified staudinger reaction. Science 287(5460):2007–2010. https://doi.org/10.1126/science.287.5460.2007

Schmid EL et al (1998) Screening ligands for membrane protein receptors by total internal reflection fluorescence: the 5-HT3 serotonin receptor. Anal Chem 70(7):1331–1338. https://doi.org/10.1021/ac9712658

Schneider AK et al (2017) DNA-SMART: biopatterned polymer film microchannels for selective immobilization of proteins and cells. Small 13(17):1603923. https://doi.org/10.1002/smll.201603923

Schumacher D et al (2018) Nanobodies: chemical functionalization strategies and intracellular applications. Angew Chem Int Ed Engl 57(9):2314–2333. https://doi.org/10.1002/anie.201708459

Seo MH et al (2014) Protein conformational dynamics dictate the binding affinity for a ligand. Nat Commun 5:3724. https://doi.org/10.1038/ncomms4724

Silva M et al (2021) Efficient amino-sulfhydryl stapling on peptides and proteins using bifunctional NHS-activated acrylamides. Angew Chem Int Ed Engl 60(19):10850–10857. https://doi.org/10.1002/anie.202016936

Slavoff SA et al (2008) Expanding the substrate tolerance of biotin ligase through exploration of enzymes from diverse species. J Am Chem Soc 130(4):1160–1162. https://doi.org/10.1021/ja076655i

Sletten EM, Bertozzi CR (2009) Bioorthogonal chemistry: fishing for selectivity in a sea of functionality. Angew Chem Int Ed Engl 48(38):6974–6998. https://doi.org/10.1002/anie.200900942

Soellner MB et al (2003) Site-specific protein immobilization by staudinger ligation. J Am Chem Soc 125(39):11790–11791. https://doi.org/10.1021/ja036712h

Staudinger H, Meyer J (1919) New organic compounds of phosphorus. II. Phosphazines. Helv Chim Acta 2:619–635. https://doi.org/10.1002/hlca.19190020163

Stenehjem ED et al (2013) Gas-phase azide functionalization of carbon. J Am Chem Soc 135(3):1110–1116. https://doi.org/10.1021/ja310410d

Sydor JR et al (2003) Chip-based analysis of protein-protein interactions by fluorescence detection and on-chip immunoprecipitation combined with microLC-MS/MS analysis. Anal Chem 75(22):6163–6170. https://doi.org/10.1021/ac034258u

Tam A et al (2007) Water-soluble phosphinothiols for traceless staudinger ligation and integration with expressed protein ligation. J Am Chem Soc 129(37):11421–11430. https://doi.org/10.1021/ja073204p

Tang L et al (2015) Bioorthogonal oxime ligation mediated in vivo cancer targeting. Chem Sci 6(4):2182–2186. https://doi.org/10.1039/c5sc00063g

Ul-Haq E et al (2013) Photocatalytic nanolithography of self-assembled monolayers and proteins. ACS Nano 7(9):7610–7618. https://doi.org/10.1021/nn402063b

Uth C et al (2014) A chemoenzymatic approach to protein immobilization onto crystalline cellulose nanoscaffolds. Angew Chem Int Ed Engl 53(46):12618–12623. https://doi.org/10.1002/anie.201404616

Vila-Perelló M (2013) Streamlined expressed protein ligation using split inteins. J Am Chem Soc 135(1):286–292. https://doi.org/10.1021/ja309126m

Wan Q, Danishefsky SJ (2007) Free-radical-based, specific desulfurization of cysteine: a powerful advance in the synthesis of polypeptides and glycopolypeptides. Angew Chem Int Ed Engl 46(48):9248–9252. https://doi.org/10.1002/anie.200704195

Wang J et al (2001a) Orientation specific immobilization of organophosphorus hydrolase on magnetic particles through gene fusion. Biomacromol 2(3):700–705. https://doi.org/10.1021/bm015517x

Wang J et al (2001b) Improving the activity of immobilized subtilisin by site-directed attachment through a genetically engineered affinity tag. Fresenius J Anal Chem 369(3–4):280–285. https://doi.org/10.1007/s002160000622

Wang T et al (2016) The pentadehydro-diels-alder reaction. Nature 532(7600):484–488. https://doi.org/10.1038/nature17429

Wang H et al (2017) Selective in vivo metabolic cell-labeling-mediated cancer targeting. Nat Chem Biol 13(4):415–424. https://doi.org/10.1038/nchembio.2297

Wang J et al (2019) Site-specific immobilization of β2-AR using O(6)-benzylguanine derivative-functionalized supporter for high-throughput receptor-targeting lead discovery. Anal Chem 91(11):7385–7393. https://doi.org/10.1021/acs.analchem.9b01268

Waugh DS (2005) Making the most of affinity tags. Trends Biotechnol 23(6):316–320. https://doi.org/10.1016/j.tibtech.2005.03.012

Weinrich D et al (2010) Oriented immobilization of farnesylated proteins by the thiol-ene reaction. Angew Chem Int Ed Engl 49(7):1252–1257. https://doi.org/10.1002/anie.200906190

Wen L et al (2018) A one-step chemoenzymatic labeling strategy for probing sialylated thomsen-friedenreich antigen. ACS Cent Sci 4(4):451–457. https://doi.org/10.1021/acscentsci.7b00573

Wingren C et al (2005) Microarrays based on affinity-tagged single-chain Fv antibodies: sensitive detection of analyte in complex proteomes. Proteomics 5(5):1281–1291. https://doi.org/10.1002/pmic.200401009

Yang CR et al (2018) Synthesis, cytotoxic evaluation and target identification of thieno[2,3-d]pyrimidine derivatives with a dithiocarbamate side chain at C2 position. Eur J Med Chem 154:324–340. https://doi.org/10.1016/j.ejmech.2018.05.028

Yu CC et al (2012) Site-specific immobilization of enzymes on magnetic nanoparticles and their use in organic synthesis. Bioconjug Chem 23(4):714–724. https://doi.org/10.1021/bc200396r

Zeng KZ et al (2018) One-step methodology for the direct covalent capture of GPCRs from complex matrices onto solid surfaces based on the bioorthogonal reaction between haloalkane dehalogenase and chloroalkanes. Chem Sci 9(2):446–456. https://doi.org/10.1039/c7sc03887a

Zhang Y et al (2015) Simultaneous site-specific dual protein labeling using protein prenyl-transferases. Bioconjug Chem 26(12):2542–2553. https://doi.org/10.1021/acs.bioconjchem.5b00553

Zhao XF et al (2020) Covalent inhibitor-based one-step method for endothelin receptor A immobilization: from ligand recognition to lead identification. Anal Chem 92(20):13750–13758. https://doi.org/10.1021/acs.analchem.0c01807

Zhu S et al (2014) Genetically encoding a light switch in an ionotropic glutamate receptor reveals subunit-specific interfaces. Proc Natl Acad Sci U S A 111(16):6081–6086. https://doi.org/10.1073/pnas.1318808111

Chapter 4
Key Biochemical Aspects of Drug-Target Interactions

Abstract The realization of drug-target interaction plays important role in many processes that determine the fate of a drug and its eventual therapeutic effects in practice. This makes it critical to fundamental research of understanding biological processes and the efforts to develop new therapeutic molecules. In this chapter, we summarize several key biochemical parameters, such as binding affinity and kinetics, dynamic conformational ensembles of drug targets, drug resistance time, rebinding, binding thermodynamics, and ligand efficiency.

Keywords Drug-target interactions · Affinity and kinetics · Thermodynamics · Ligand efficiency

Abbreviations

k_{on}	Rate constants of association
k_{off}	Rate constants of dissociation
K_D	Equilibrium dissociation constant
AT_1R	Angiotensin II type 1 receptor
$\Delta G°$	Gibbs standard free binding energy
$\Delta H°$	Equilibrium enthalpy
$\Delta S°$	Entropy

As the most predominant family of drug targets, the studies on drugs binding to GPCRs start with the receptor theory of Ehrlich (Tait 1913) and the contemporaneous lock-and-key theory of Fischer (Emil 1894). During this period, drug-receptor binding interactions have been commonly viewed that the binding is driven by structural complementarity between the drug and the protein. In recent decades, advances in biotechnology and pharmacology have highlighted the importance of several biochemical aspects in mediating drug-target binding. These aspects are summarized as binding affinity and kinetics, target conformational ensembles, drug resistance time, rebinding, binding thermodynamics, binding kinetics, and ligand efficiency.

© The Author(s), under exclusive license to Springer Nature Singapore Pte Ltd. 2023
X. Zhao et al., *G Protein-Coupled Receptors*, SpringerBriefs in Molecular Science,
https://doi.org/10.1007/978-981-99-0078-7_4

4.1   Binding Affinity and Kinetics

The affinity of a drug binding to a specific receptor behaves as a key step that determines the in vivo efficacy of a drug. It is usually assessed by the equilibrium dissociation constant (K_D) in a closed system. However, drug-receptor interactions are often far away from equilibrium conditions in the human body since dynamic flow and drug metabolism prevents the achievement of equilibrium (Gleeson et al. 2011). This promotes the postulations about assessing in vivo drug effects by binding kinetics in terms of rate constants of association (k_{on}) and dissociation (k_{off}). The ratio of the two parameters is theoretically equal to the equilibrium dissociation constant ($K_D = k_{off}/k_{on}$) (Sykes et al. 2019).

4.2   Dynamic Conformational Ensembles of Drug Targets

An enormous amount of research has demonstrated that GPCR dynamic conformations play important roles in both the association and dissociation processes even for many high-affinity and high-efficacy drugs. Since it is impossible to capture all these dynamic conformations, conventional determination of binding affinity and kinetics assumes a single and stable receptor conformation under certain conditions. This assumption is far distinct from the reality that GPCR adopts multiple conformations on the cell surface and the proof of favorable conformations for a ligand binding to the receptor (Boehr et al. 2009). A case study of β_2-AR has confirmed that a structural approach is indeed biased toward finding inverse agonists (Rai et al. 2009). These results necessitate the consideration of conformational ensembles during drug-receptor interaction analysis.

4.3   Drug Resistance Time

The concept of drug-target resistance time and its impact on drug discovery has been well-reviewed by Copeland (2010). This concept is proposed to describe experimental measurements that are related to the lifetime of the binary drug-target complex, and this in turn to durable, pharmacodynamic activity (Swinney 2004). In standard experiments, the drug-target residence time represents the time to reach around 63% dissociation of the drug-target complex. This definition is elicited by the study of tiotropium. The drug demonstrates approximately equal binding affinity to both muscarinic M2 and M3 receptors while exhibiting a dissociation rate constant 10 times faster from the M2 receptor than the M3 receptor. As the primary target distributed airway tissues, the M3 receptor mediates the main efficacy of the drug, whereas the M2 receptor induces its side effects on the cardiovascular system. This indicates that a faster dissociation rate is preferable when the drug's adverse effect mainly comes from the drug-receptor complex itself. This is the designing strategy of atypical neuroleptics like clozapine which dissociates more rapidly from its primary target dopamine D2 receptor than the first-generation neuroleptics with a slow off rate. Owing to the faster dissociation rate, these drugs prove to have less extrapyramidal side effects such as Parkinsonian-like syndrome and tardive dyskinesia (Tummino and Copeland 2008).

4.4 Rebinding

Rebinding is widely accepted as an alternative for improving the duration of drug action, besides long-lasting target binding (Vauquelin 2010). At initial, pharmacologists define rebinding as the phenomenon that occurs when the free ligands are not removed by wash-out experiments or when their free concentration is not completely reduced in radioligand binding studies. Subsequent theoretical and experimental attempts move this definition to the capacity of a drug repeatedly binding to the same receptor prior to being diffused away from the target. This implies that the rebinding is determined by the association rate rather than by the affinity of the ligand-receptor complex. Inspired by the above definition and implication, previous reports have demonstrated that rebinding takes place for several antagonists when they interact with their specific targets such as the angiotensin II type 1 receptor (AT_1R) and the dopamine D2 receptor (Vauquelin and Charlton 2010).

4.5 Binding Thermodynamics

The typical receptor binding assays determine the binding affinity of drug-receptor interaction at a single temperature. This is unable to provide sufficient information for realizing the molecular mechanisms underlying the interaction. As an alternative, receptor binding thermodynamics is a powerful tool to gain deep insight, at the molecular level, of the events that occur during drug-receptor interactions. In the thermodynamic study, the drug-receptor binding mechanism is characterized by the molecular forces that drive the interaction. These forces are determined by classical parameters such as Gibbs standard free binding energy ($\Delta G°$), the equilibrium enthalpy ($\Delta H°$), and entropy ($\Delta S°$). As defined by the Gibbs equation, $\Delta G° = \Delta H° - T\Delta S°$. According to semiempirical law by Ross and Subramanian (1981), the forces are typically considered as hydrogen bonds, van der Waals forces, and hydrophobic interactions. Numerous studies have confirmed that the optimization of drug-receptor binding enthalpy benefits the achievement of medicinal chemistry optimization purposes (Sykes et al. 2019).

4.6 Ligand Efficiency

Ligand efficiency has been first defined as the ratio of ΔG to the number of non-hydrogen atoms of the ligand for medicinal chemistry optimization (Hopkins et al. 2004). This provides an interesting strategy for drug design, in which the concept has been described as several derivatives such as the binding efficiency index (Polanski and Tkocz 2017), percentage efficiency, and surface binding efficiency index (Abad-Zapatero et al. 2010), kinetic efficiency (Holdgate and Gill 2011) and lipophilic ligand efficiency (Hopkins et al. 2014). These reports have implied that both ligand size and the binding sites of the receptor are decisive in ligand efficiency. In most cases, smaller ligands have greater efficiencies than larger ligands given that the receptor remains identical. This makes ligand efficiency an attractive and powerful strategy in the practice of medicinal chemistry.

References

Abad-Zapatero C et al (2010) Ligand efficiency indices for an effective mapping of chemico-biological space: the concept of an atlas-like representation. Drug Discov Today 15(19–20):804–811. https://doi.org/10.1016/j.drudis.2010.08.004

Boehr DD et al (2009) The role of dynamic conformational ensembles in biomolecular recognition. Nat Chem Biol 5(11):789–796. https://doi.org/10.1038/nchembio1209-954d

Copeland RA (2010) The dynamics of drug-target interactions: drug-target residence time and its impact on efficacy and safety. Expert Opin Drug Discov 5(4):305–310. https://doi.org/10.1517/17460441003677725

Emil F (1894) Einfluss der configuration auf die wirkung der enzyme. II. Berichte Der Deutschen Chemischen Gesellschaft. https://doi.org/10.1002/cber.189402703169

Gleeson MP et al (2011) Probing the links between in vitro potency, ADMET and physicochemical parameters. Nat Rev Drug Discov 10(3):197–208. https://doi.org/10.1038/nrd3367

Holdgate GA, Gill AL (2011) Kinetic efficiency: the missing metric for enhancing compound quality? Drug Discov Today 16(21–22):910–913. https://doi.org/10.1016/j.drudis.2011.09.006

Hopkins AL et al (2004) Ligand efficiency: a useful metric for lead selection. Drug Discov Today 9(10):430–431. https://doi.org/10.1016/S1359-6446(04)03069-7

Hopkins AL et al (2014) The role of ligand efficiency metrics in drug discovery. Nat Rev Drug Discov 13(2):105–121. https://doi.org/10.1038/nrd4163

Polanski J, Tkocz A (2017) Between descriptors and properties: understanding the ligand efficiency trends for G protein-coupled receptor and kinase structure-activity data sets. J Chem Inf Model 57(6):1321–1329. https://doi.org/10.1021/acs.jcim.7b00116

Rai P et al (2009) Continuous elimination of oxidized nucleotides is necessary to prevent rapid onset of cellular senescence. Proc Natl Acad Sci U S A 106(1):169–174. https://doi.org/10.1073/pnas.0809834106

Ross PD, Subramanian S (1981) Thermodynamics of protein association reactions: forces contributing to stability. Biochemistry 20(11):3096–3102

Swinney DC (2004) Biochemical mechanisms of drug action: what does it take for success? Nat Rev Drug Discov 3(9):801–808. https://doi.org/10.1038/nrd1500

Sykes DA et al (2019) Binding kinetics of ligands acting at GPCRs. Mol Cell Endocrinol 485:9–19. https://doi.org/10.1016/j.mce.2019.01.018

Tait A (1913) Address in pathology on chemotherapeutics. Lancet 182(4694):445–451

Tummino PJ, Copeland RA (2008) Residence time of receptor-ligand complexes and its effect on biological function. Biochemistry 47(20):5481–5492. https://doi.org/10.1021/bi8002023

Vauquelin G (2010) Rebinding: or why drugs may act longer in vivo than expected from their in vitro target residence time. Expert Opin Drug Discov 5(10):927–941. https://doi.org/10.1021/bi8002023

Vauquelin G, Charlton SJ (2010) Long-lasting largest binding and rebinding as mechanisms to prolong in vivo drug action. Br J Pharmacol 161(3):488–508. https://doi.org/10.1111/j.1476-5381.2010.00936.x

Chapter 5
Immobilized GPCRs in Drug-Receptor Interaction Analysis

Abstract The development of new assays for understanding GPCR-drug interaction has witnessed many breakthroughs during the last decades. Since then, a number of additional examples of new approaches and/or applications for decoding the interaction between drugs and GPCRs have appeared. In this chapter, we present some of the reports using immobilized GPCRs for drug-receptor interaction analysis, including labelled and label-free assays.

Keywords Labelled drug-receptor binding assays · Label-free drug-receptor binding assay

Abbreviations

BRET	Bioluminescence resonance energy transfer
RWG	Resonant waveguide grating
HPLC	High-performance liquid chromatography
FAC	Frontal affinity chromatography
MS	Mass spectra
AED	Adsorption energy distribution
ET_AR	Endothelin A receptor
NLC	Nonlinear chromatography
K_A	Association constant

5.1 Labeled Drug-Receptor Binding Assays

5.1.1 Radioligand Binding Assays

The radioligand-binding assay has been implemented in the mid-twentieth century. It allows direct characterization of the binding interaction between drugs to their receptors in any matrices consisting of a purified receptor, cells, and tissues. The main disadvantages of radioligand-binding assay lay behind several aspects: the hard work for synthesizing a radioligand; the need for removal of non-bound ligands and the management of radioactive waste. Despite these drawbacks, this method has remained the most popular alternative for decades, particularly when it comes to the binding of drugs to GPCRs. This is reasoned by the difficulty and cost to achieve functional GPCRs with high purification which are necessary for other binding assays.

Conventionally, radioligand binding assays have been performed in terms of saturation assay, indirect assay, and kinetic assay (Flanagan 2016). The three types of protocols can determine the receptor density on a surface, the drug affinity to the receptor, and the dissociation rate constant of the interaction. Recent reports have applied radioligand-binding assay in measuring ligand binding to GPCRs on the cell surface, for instance, α_{2B}-adrenergic receptor (Wei et al. 2019) and dopamine receptors (Bada et al. 2020).

5.1.2 Fluorescent Ligand Binding Assays

Fluorescence methods have emerged as a promising player in characterizing the binding of ligands to GPCRs due to the successful synthesis of diverse fluorescent dyes. As highlighted by several reviews, recent technological breakthroughs enable these methods to determine all the aforementioned biochemical properties of the interaction between drugs and GPCRs, ranging from equilibrium-binding parameters to kinetic and thermodynamic parameters (Iliopoulos-Tsoutsouvas et al. 2018; Rinken et al. 2018; Stoddart et al. 2016).

Among these techniques, one approach termed bioluminescence resonance energy transfer (BRET) assay has been developed by Hill and his colleagues, allowing them to monitor binding to the human β_2-adenoceptor (β_2-AR) in real-time using a red fluorescent analog of the diverse antagonist. By using an N-terminal NanoLuc tag, they further improved the method as NanoBRET and extend the application in investigating the binding of distinct fluorescent ligands to the human β_1-adenoceptor (β_1-AR) expressed in HEK-293 cells (Soave et al. 2016) and to the human β_2-AR in a mouse model of breast cancer (Alcobia et al. 2018). Subsequent work by them and other scientists have further strengthened the utility of this method for a range of applications, especially in GPCR binding fields (Kozielewicz et al. 2020; Comeo et al. 2020; White et al. 2019). The main drawback of this method is likely the difficulty of obtaining good fluorescent ligands and the alterations in binding characteristics of the original drugs due to the introduction of bulky fluorescent tags.

5.2 Label-Free Drug-Receptor Binding Assays

5.2.1 Surface Plasmon Resonance (SPR)

It is widely accepted that most of the label-dependent assays are partly laborious and prone to artifacts due to the requirements of modifying at least one of the binding partners when they are applied in the drug-receptor interaction analysis. This allows label-free binding techniques to draw a splendid figure for understanding the binding of drugs to GPCRs in diverse aspects of the field. Among these techniques, surface plasmon resonance is the most popular approach for detecting the binding of ligands to the receptors immobilized on a gold surface.

Surface plasmon resonance-related methods have proved to be good choices for distinct protein classes, particularly soluble proteins like enzymes, kinases, and proteases. Unlike these applications, the utilization of such assays in GPCRs remains hard to advance because of the difficulty to purify GPCRs and functionally reconstitute them into lipid environments to pursue reproducible and robust signals. In this context, successful cases have been reviewed in the characterization of ligands binding to rhodopsin, chemokine receptors CCR5 and CXCR4, neurotensin receptor-1, olfactory receptor, neuropeptide Y4 receptor, adenosine-A2A receptor and β_1-adrenergic receptor (Patching 2014). Other earlier reports have demonstrated that the technique called plasmon-waveguide resonance technology enables the examination of conformational changes during δ-opioid receptor-ligand interaction (Alves et al. 2004). Recent studies include the kinetic and thermodynamic characterization of opioids interaction with the human μ-opioid receptor (Babazada et al. 2019) and the investigation of ligand binding of the engineered GPR17 receptor (Capelli et al. 2019). These cases are important in paving a way for the broad application of the method in the GPCR field.

With the advances in instrumentation, recent researchers have developed an exciting new technique named surface plasmon resonance microscopy, which enables measurement of binding kinetics of membrane proteins from single-molecule sensing to single-cell imaging (Zhou et al. 2020a, b). This method has been successfully applied in mapping the local distribution and binding activities of nicotinic acetylcholine receptors in single living cells (Wang et al. 2012). Recently, this assay has also been employed to explore the pharmacodynamic responses of neuronal cells expressing the muscarinic acetylcholine receptor to nerve growth factor and muscarine, a nonselective agonist of the receptor (Mir and Shinohara 2017). The two cases have demonstrated that surface plasmon resonance microscopy is possible to monitor the binding of ligands to receptors on the cell surface with millisecond temporal and micrometer spatial resolution. Besides, it has substantial enhancement in sensitivity over conventional surface plasmon resonance assays since it allows measurement of drug-receptor binding from intensive sample pots. These advantages are expected to result in increasing application of the method in understanding binding mechanisms of ligands to GPCRs as improvements are made to purify and stabilize the receptors.

5.2.2 Resonant Waveguide Grating (RWG)

Similar to surface plasmon resonance, the surface-bound electromagnetic wave has also been utilized for biomolecular analysis in another method termed resonant waveguide grafting, which is often reported as a grating coupler or photonic crystal biosensor in the literature. At initial, such kinds of methods are devised to determine drug-receptor binding affinity, rather than the examination of other parameters like kinetics, thus limits their application in profiling the interaction between ligands and GPCRs.

Recent reports by Wells et al. have introduced resonant waveguide grafting biosensors into screening agonists and antagonists that regulate the interaction between Gbg protein and soluble N-ethylmaleimide-sensitive factor attachment proteins. Among these proteins, several GPCRs such as dopamine D2 receptor, 5HT1 serotonin receptors, and M4 muscarinic receptor are well examined (Wells et al. 2012). Most recently, Liang et al. have applied this method in phenotypic assessment and ligand screening of ETA/ETB receptors by measuring protein mass changes associated with GPCR-mediated signaling events, a phenomenon which is known as dynamic mass redistribution. Similar strategies have also been employed for the analysis of endogenous adrenergic receptor signaling in primary preadipocytes and differentiated adipocytes (Jørgensen et al. 2019), identification of novel phytocannabinoids from Ganoderma (Zhou et al. 2020a, b), and systematic characterization of AT1 receptor (AT_1R) antagonists (Qu et al. 2020). These breakthroughs highlight the allowance of a sophisticated approach to more specialized scientific questions about ligand mediating GPCR functions using the improved technique of resonant waveguide grafting.

5.2.3 Interferometry Biosensor

Interferometry biosensor, typically used as biolayer interferometry, is a well-established optical technique to study drug-receptor interaction at the liquid–solid surface (Lee et al. 2019; Janezic et al. 2019). This method applies a spectrometer to capture the interference pattern of visible light reflected from two surfaces on which an immobilized receptor layer and an internal reference are formed. This interference pattern depends on the binding of ligands to the receptors, thus produces an intensity variation with a characteristic profile of peaks and troughs. The conventional format of bilayer interferometry is often based on a disposable sensor head in the presence of an array of 8 or 16 sensors that are immersed in 96-well or 384-well plates. This allows the simultaneous determination of ligand binding and dissociation, thus makes this technique high throughput for binding kinetics (Wartchow et al. 2011).

Recent representative examples for the application of biolayer interferometry in GPCR fields include the measurement of ligand efficacy at the μ-opioid receptor (Livingston et al. 2018) and the characterization of the β_2-AR stimulation with dopamine (Verzijl et al. 2017). These reports make biolayer interferometry advantageous options for determining the kinetics of ligands binding to GPCR. While biolayer interferometry is possible to use cell membranes for GPCR study, substantial noise usually masks the signal when the ligands are weak binders of the receptor. This problem is often alleviated by using purified receptors. Owning to the great progress in protein engineering and immobilization strategy, biolayer interferometry has the potential to become a powerful alternative for quantification of interactions between ligands and GPCRs.

5.2.4 Atom Force Microscopy

Since invented by Binnig et al. in 1986, atomic force microscopy has attracted continuous attention from diverse fields that make it a multifaceted tool for studying proteins and their interactions with ligands on certain surfaces (Binnig et al. 1986). Nowadays, several reports have reviewed the establishments and applications of recent new atomic force microscopy-based methods in distinct aspects, especially their capabilities for membrane protein studies (Alsteens et al. 2017; Whited and Park 2014).

Among the recent reports, a few works have introduced atomic force microscopy in quantifying the physical and chemical properties of diverse GPCR conformations and sensing their interactions with ligands. These applications are summarized as: directly observing the supramolecular assembly of GPCRs in native membranes (Aguzzi and Heikenwalder 2003), profiling the free-energy landscape of the ligands binding to GPCRs (Alsteens et al. 2015), and realizing how the interactions modulate the GPCR structural properties (Puntheeranurak et al. 2011). Relatively new reviews (Sapra et al. 2019) also highlighted the significance and applications of atomic force microscopy-based methods in localizing ligand-binding events on single GPCRs, studying changes in interactions between receptors and ligands, other molecules or proteins mediating GPCR functions. Despite these dramatic advances, atomic force microscopy is still challenged when it comes to determining the stability of GPCRs in complex with G-proteins and arrestin. Thus, the future focus of the methods is highlighted to address how ligand-binding events modulate GPCR states, and to correlate the information with the data provided by functional assays.

5.2.5 Receptor Chromatography (Column-Based Assays)

Receptor chromatography is defined as a liquid chromatographic technique that makes use of the biological interaction between a receptor and its ligands for the separation and analysis of a specific analyte within a sample (Schiel et al. 2010; Hage 1999). In this method, purified GPCRs or membranes containing GPCRs are immobilized on small, rigid, and porous supports. These supports are silica gels or synthetic polymers that can be used in high-performance liquid chromatography (HPLC). The retentions of the analytes on the receptor chromatographic column are based on the specific and reversible interactions that are found in biological systems. Thus, receptor chromatography is very popular in drug-receptor interaction analysis and bioactive compound screening studies.

5.2.5.1 Frontal Chromatography

Frontal affinity chromatography (FAC) is the most accurate chromatographic method for protein–ligand interaction analysis, which is developed by Kasai and Ishii in 1975 (Kasai and Ishii 1975). In this technique, ligands flow through the affinity column and binding with the immobilized receptors. Then the characteristic breakthrough curves are formed based on the ligand affinities with the receptor. Both the low-throughput interaction characterization (i.e. K_D determination) and high throughput screening (i.e. ligands ranking) can be achieved by using any detector appropriate for the ligands, such as UV–Vis, fluorescence, radio-detector, and mass spectra (MS). FAC does not separate compounds, so the detection of multiple bindings requires a detector capable of discriminating among coeluting ligands. In this case, frontal affinity chromatography coupled to mass spectrometry (FAC-MS) is considered as the most flexible and generalized strategy for protein–ligand interaction analysis (Calleri et al. 2009).

FAC has been successfully applied to a wide range of GPCRs in many pharmacological studies, within which the binding constant measurement and screening studies are the two most prevalent works. By FAC, it is possible to determine the affinity constant between an immobilized receptor and a ligand by continuously flushing the ligand at different concentrations through the affinity stationary phase. The adsorbed amount (q^*) at equilibrium at a given ligand concentration (C) can be obtained by Eq. (5.1) (Juang and Ju 1997):

$$q^* = \frac{[C](V_{eq} - V_0)}{V_a} = [C](t_R - t_0)v_C \tag{5.1}$$

where V_{eq}, V_0 and V_a are the breakthrough volume, the void volume and the volume of stationary phase, respectively; t_R and t_0 represent the breakthrough time of the ligands and the void time of the system; v_C is the flow rate.

The mathematical model behind FAC is based on the linearization of the Langmuir equation. However, the idealized linearity rarely exists in practical cases. To improve the accuracy of the adsorption model selection, we introduced adsorption energy distribution (AED), a numerical tool based on the expectation–maximization algorism, to study the β_2-AR-ligand and endothelin A receptor (ET$_A$R)-ligand interactions (Li et al. 2018). The extension of AED to FAC is the potential for the precise and high-throughput clarification of GPCR-drug interactions.

The binding of multiple ligands to the immobilized receptor in a simultaneous study can be accomplished using detection at selected m/z values by FAC-MS. In the presence of an indicator, the method allows ranking the affinity of ligand mixtures to the immobilized receptor through Eq. (5.2) (Slon-Usakiewicz et al. 2005):

$$Shift\ (\%) = \frac{t_1 - t}{t_1 - t_0} \times 100\% \tag{5.2}$$

where t_1 and t represent the breakthrough times of the indicator in the absence or presence of the ligand. The greater the percentage shifts, the greater the degree of competition for the indicator. This method has been successfully applied in finding active compounds from traditional Chinese herbs (Luo et al. 2003).

5.2.5.2 Zonal Elution

Zonal elution is another prevalent way for GPCR-ligand interaction in receptor chromatography, which is usually applied for ligand competition experiments. In this method, a small amount of ligand is injected into an affinity column with immobilized GPCR. The injection is often made in the presence of a mobile phase containing a known concentration of a competing ligand. By varying the concentration of the competing ligand (I), the retention time of the injected ligand (A) is monitored and used to provide the binding information between the ligands and immobilized GPCR through the equations as follows (Li et al. 2014):

$$\frac{1}{k' - X} = \frac{V_m K_I [I]}{K_A m_L} + \frac{V_m}{K_A m_L} \tag{5.3}$$

$$\frac{1}{\sqrt{k'}} = \frac{K_A}{\sqrt{K_A m_L}} [A] + \frac{1}{\sqrt{K_A m_L}} \tag{5.4}$$

where K_A and K_I are the association constant of the injected and competing ligands with the immobilized receptor; m_L is the numbers of binding sites; V_m is the void volume of the chromatographic system; X is the capacity factor caused by the nonspecific interaction on the column. If the competing and injected ligands are the same, Eq. (5.4) is used for binding constant calculation; otherwise Eq. (5.3) should be applied.

The advantages of zonal elution are as follows: (i) a small amount of ligand is needed in each injection; (ii) more than one ligand (e.g. chiral ligands) per injection can be examined (Hage et al. 2009); (iii) a ligand with multiple sites to the immobilized receptor can be investigated at individual sites (Matsuda et al. 2015).

5.2.5.3 Nonlinear Chromatography

Nonlinear chromatography (NLC) was first derived by Thomas (1944) and extended to solve the impulse input in affinity chromatography by Wade in 1987 (Wade et al.

1987). Due to the fast adsorption and slow desorption process in affinity chromatography, the impulse input often follows non-Gaussian peaks with an extremely tailing profile. The tailed chromatogram can be input into the software PeakFit for signal deconvolution through Eqs. (5.5) and (5.6):

$$y = \frac{a0}{a3}\left[1 - \exp\left(-\frac{a3}{a2}\right)\right]\left[\frac{\sqrt{\frac{a1}{x}}I_1\left(\frac{2\sqrt{a1x}}{a2}\right)\exp\left(\frac{-x-a1}{a2}\right)}{1 - T\left(\frac{a1}{a2}, \frac{x}{a2}\right)\left[1 - \exp\left(-\frac{a3}{a2}\right)\right]}\right] \tag{5.5}$$

$$T(u, v) = \exp(-v)\int_0^u \exp(-t)I_0\left(2\sqrt{vt}\right)dt \tag{5.6}$$

where x and y are the reduced retention time and signal intensity; $I_0()$ and $I_1()$ are Modified Bessel functions; a_0, a_1, a_2, and a_3 are the area, thermodynamic, width, and distortion parameters, respectively. C_0 is the concentration of the solute injected multiplied by the width of the injection pulse. The adsorption/desorption rate constant (k_d/k_a) and association constant (K_A) can be obtained by the equations: $k_d = 1/a_2t_0$; $k_a = k_dK_A$; $K_A = a_3/C_0$.

There are several advantages to use NLC for receptor-ligand interaction analysis: (i) the injection amount of the ligand is very small because the hypothesis of NLC is valid only at infinite dilution; (ii) NLC can be used for the determination of the equilibrium and rate constants simultaneously in one injection; (iii) NLC is still valid when the tailing of the chromatogram profile is extremely serious.

5.2.5.4   Injection Amount-Dependent Method

Injection amount dependent analysis is a simple mathematical model for GPCR-drug interaction analysis, which is proposed by our group in 2014 (Zhao et al. 2014). It assumes that the adsorption sites are evenly distributed on the immobilized receptor and the equilibration is a rapid process with negligible longitudinal diffusion. The association constant (K_A) of a ligand binding to an immobilized receptor can be illustrated by Eq. (5.7)

$$\frac{k'}{1 + k'} = n_t - \frac{1}{K_A} \times k'V_m \tag{5.7}$$

where n_t is the total molar amount of adsorption sites. The validation of the method has been proved by the interaction between ligands and immobilized proteins, such as human serum albumin, β_2-AR, ET_AR and AT_1R (Sun et al. 2019; Zeng et al. 2017). The method avoids saturating the affinity column with a large number of ligands and has the advantages of ligand- and time-saving.

5.2.5.5 Adsorption Energy Distribution

Adsorption energy distribution (AED) is an important model to determine the degree of heterogeneity in interactions (Samuelsson et al. 2009). The superiority of AED is the selection of the correct adsorption model without a model assumption beforehand. Taking the Langmuir isotherm as the local model, one can obtain (Fritti et al. 2003):

$$q^*([C]) = \int_0^\infty F(\varepsilon) \left(\frac{b_0 e^{\varepsilon/RT}[C]}{1 + b_0 e^{\varepsilon/RT}[C]} \right) d\varepsilon \tag{5.8}$$

The distribution function $F(\varepsilon)$ is calculated by the expectation maximization method which is directly applied to the raw data. This method allows a robust AED estimate, minimizes artifactual information from a numerical standpoint and provides the least biased solution available. $F(\varepsilon)$ is discretized using N grid points in the energy space between ε_{min} and ε_{max}. The amount $q([C_j])$ at concentration $[C_j]$ is iteratively determined by the following equation:

$$q_{cal}^k([C_j]) = \sum_{i=1}^{i=N} F^k(\varepsilon_i) \cdot \frac{b_0 e^{\varepsilon_i/RT}[C_j]}{1 + b_0 e^{\varepsilon_i/RT}[C_j]} \cdot \Delta\varepsilon \quad j \in [1, M]; i \in [1, N] \tag{5.9}$$

where $\Delta\varepsilon$ and ε_i are defined by the following equations:

$$\Delta\varepsilon = \frac{\varepsilon_{max} - \varepsilon_{min}}{N - 1} \tag{5.10}$$

$$\varepsilon_i = \varepsilon_{min} + (i - 1)\Delta\varepsilon \tag{5.11}$$

The initial expression of AED function is as follows:

$$F^0(\varepsilon_i) = \frac{q(C_M)}{N} \quad \forall i \in [1, N] \tag{5.12}$$

M, N, b_{min}, b_{max} and the iteration numbers are defined to begin an AED calculation, and the final result is a distribution of equilibrium constants. Then the AED function at the kth iteration can be updated as Eq. (5.13):

$$F^{k+1}(\varepsilon_i) = F^k(\varepsilon_i) \sum_{j=1}^{j=M} \frac{b_0 e^{\varepsilon_i/RT}[C_j]}{1 + b_0 e^{\varepsilon_i/RT}[C_j]} \cdot \Delta\varepsilon \frac{q_{exp}([C_j])}{q_{cal}^k([C_j])} \tag{5.13}$$

The energy distributions for different adsorption models are as follows: (i) the Langmuir model: symmetrical, unimodal energy distribution; (ii) the BiLangmuir model: bimodal energy distribution; (iii) the Tóth model: asymmetrical, unimodal energy distribution with tail towards the low adsorption energies (Fritti et al. 2003).

References

Aguzzi A, Heikenwalder M (2003) Prion diseases: cannibals and garbage piles. Nature 423(6936):127–129. https://doi.org/10.1038/423127a

Alcobia DC et al (2018) Visualizing ligand binding to a GPCR in vivo using NanoBRET. iScience 6:280–288. https://doi.org/10.1016/j.isci.2018.08.006

Alsteens D et al (2015) Imaging G protein-coupled receptors while quantifying their ligand-binding free-energy landscape. Nat Methods 12(9):845–851. https://doi.org/10.1038/NMETH.3479

Alsteens D et al (2017) Atomic force microscopy-based characterization and design of biointerfaces. Nat Rev Mater 2:17008. https://doi.org/10.1038/natrevmats.2017.8

Alves ID et al (2004) Different structural states of the proteolipid membrane are produced by ligand binding to the human delta-opioid receptor as shown by plasmon-waveguide resonance spectroscopy. Mol Pharmacol 65(5):1248–1257. https://doi.org/10.1124/mol.65.5.1248

Babazada H et al (2019) Biosensor-based kinetic and thermodynamic characterization of opioids interaction with human μ-opioid receptor. Eur J Pharm Sci 138:105017. https://doi.org/10.1016/j.ejps.2019.105017

Bada J et al (2020) Detergent-free extraction of a functional low-expressing GPCR from a human cell line. Biochim Biophys Acta Biomembr 1862(3):183152. https://doi.org/10.1016/j.bbamem.2019.183152

Binnig G et al (1986) Atomic force microscope. Phys Rev Lett 56(9):930–933. https://doi.org/10.1103/PhysRevLett.56.930

Calleri E et al (2009) Target-based drug discovery: the emerging success of frontal affinity chromatography coupled to mass spectrometry. ChemMedChem 4(6):905–916. https://doi.org/10.1002/cmdc.200800436

Capelli D et al (2019) Surface plasmon resonance as a tool for ligand binding investigation of engineered GPR17 receptor, a G protein coupled receptor involved in myelination. Front Chem 7:910. https://doi.org/10.3389/fchem.2019.00910

Comeo E et al (2020) Subtype-selective fluorescent ligands as pharmacological research tools for the human adenosine A(2A) receptor. J Med Chem 63(5):2656–2672. https://doi.org/10.1021/acs.jmedchem.9b01856

Flanagan CA (2016) GPCR-radioligand binding assays. Methods Cell Biol 132:191–215. https://doi.org/10.1016/bs.mcb.2015.11.004

Fritti F et al (2003) Determination of single component isotherms and affinity energy distribution by chromatography. J Chromatogr A 988:185–203. https://doi.org/10.1016/S0021-9673(02)02084-8

Hage DS (1999) Affinity chromatography: a review of clinical applications. Clin Chem 45(5):593–615

Hage DS et al (2009) Characterization of drug-protein interactions in blood using high-performance affinity chromatography. J Sep Sci 32(5–6):835–853. https://doi.org/10.1002/jssc.200800640

Iliopoulos-Tsoutsouvas C et al (2018) Fluorescent erobes for G-protein-coupled receptor drug discovery. Expert Opin Drug Discov 13(10):933–947. https://doi.org/10.1080/17460441.2018.1518975

Janezic EM et al (2019) Scribble co-operatively binds multiple α(1D)-adrenergic receptor C-terminal PDZ ligands. Sci Rep 9(1):14073. https://doi.org/10.1038/s41598-019-50671-6

Jørgensen CV et al (2019) Label-free dynamic mass redistribution analysis of endogenous adrenergic receptor signaling in primary preadipocytes and differentiated adipocytes. J Pharmacol Toxicol Methods 97:59–66. https://doi.org/10.1016/j.vascn.2019.03.005

Juang RS, Ju CY (1997) Equilibrium sorption of copper(II)-ethylenediaminetetraacetic acid chelates onto cross-linked, polyaminated chitosan beads. Ind Eng Chem Res 36(12):5403–5409. https://doi.org/10.1021/ie970322k

Kasai K, Ishii S (1975) Quantitative analysis of affinity chromatography of trypsin. A new technique for investigation of protein-ligand interaction. J Biochem 77(1?):261–264

Kozielewicz P et al (2020) A NanoBRET-based binding assay for smoothened allows real-time analysis of ligand binding and distinction of two binding sites for BODIPY-cyclopamine. Mol Pharmacol 97(1):23–34. https://doi.org/10.1124/mol.119.118158

Lee AW et al (2019) A knottin scaffold directs the CXC-chemokine-binding specificity of tick evasins. J Biol Chem 294(29):11199–11212. https://doi.org/10.1074/jbc.RA119.008817

Li Q et al (2014) Immobilised histidine tagged β2-adrenoceptor oriented by a diazonium salt reaction and its application in exploring drug-protein interaction using ephedrine and pseudoephedrine as probes. PLoS ONE 9(4):e94955. https://doi.org/10.1371/journal.pone.0094955

Li Q et al (2018) Reliable analysis of the interaction between specific ligands and immobilized beta-2-adrenoceptor by adsorption energy distribution. Anal Chem 90(13):7903–7911. https://doi.org/10.1021/acs.analchem.8b00214

Livingston KE et al (2018) Measuring ligand efficacy at the mu-opioid receptor using a conformational biosensor. Elife 7:e32499. https://doi.org/10.7554/eLife.32499

Luo H et al (2003) Frontal immunoaffinity chromatography with mass spectrometric detection: a method for finding active compounds from traditional Chinese herbs. Anal Chem 75(16):3994–3998. https://doi.org/10.1021/ac034190i

Matsuda R et al (2015) Analysis of multi-site drug-protein interactions by high-performance affinity chromatography: binding by glimepiride to normal or glycated human serum albumin. J Chromatogr A 1408:133–144. https://doi.org/10.1016/j.chroma.2015.07.012

Mir TA, Shinohara H (2017) Two-dimensional surface plasmon resonance imaging system for cellular analysis. Methods Mol Biol 1571:31–46. https://doi.org/10.1007/978-1-4939-6848-0_3

Patching SG (2014) Surface plasmon resonance spectroscopy for characterisation of membrane protein-ligand interactions and its potential for drug discovery. Biochim Biophys Acta 1838(1PtA):43–55. https://doi.org/10.1016/j.bbamem.2013.04.028

Puntheeranurak T et al (2011) Single-molecule recognition force spectroscopy of transmembrane transporters on living cells. Nat Protoc 6(9):1443–1452. https://doi.org/10.1038/nprot.2011.370

Qu L et al (2020) Systematic characterization of AT1 receptor antagonists with label-free dynamic mass redistribution assays. J Pharmacol Toxicol Methods 102:106682. https://doi.org/10.1016/j.vascn.2020.106682

Rinken A et al (2018) Assays with detection of fluorescence anisotropy: challenges and possibilities for characterizing ligand binding to GPCRs. Trends Pharmacol Sci 39(2):187–199. https://doi.org/10.1016/j.tips.2017.10.004

Samuelsson J, Arnell R, Fornstedt T (2009) Potential of adsorption isotherm measurements for closer elucidating of binding in chiral liquid chromatographic phase systems. J Sep Sci 32:1491–1506. https://doi.org/10.1002/jssc.200900165

Sapra KT et al (2019) Seeing and sensing single G protein-coupled receptors by atomic force microscopy. Curr Opin Cell Biol 57:25–32. https://doi.org/10.1016/j.ceb.2018.10.006

Schiel JE et al (2010) Biointeraction affinity chromatography: general principles and recent developments. Adv Chromatogr 48:145–193

Slon-Usakiewicz JJ et al (2005) Global kinase screening. Applications of frontal affinity chromatography coupled to mass spectrometry in drug discovery. Anal Chem 77(5):1268–1274. https://doi.org/10.1021/ac048716q

Soave M et al (2016) Use of a new proximity assay (NanoBRET) to investigate the ligand-binding characteristics of three fluorescent ligands to the human β(1)-adrenoceptor expressed in HEK-293 cells. Pharmacol Res Perspect 4(5):e00250. https://doi.org/10.1002/prp2.250

Stoddart LA et al (2016) Fluorescence- and bioluminescence-based approaches to study GPCR ligand binding. Br J Pharmacol 173(20):3028–3037. https://doi.org/10.1111/bph.13316

Sun H et al (2019) Characterization of the heterogeneous adsorption of three drugs on immobilized bovine serum albumin by adsorption energy distribution. J Chromatogr B Analyt Technol Biomed Life Sci 1125:121727. https://doi.org/10.1016/j.jchromb.2019.121727

Thomas HC (1944) Heterogeneous ion exchange in a flowing system. J Am Chem Soc 66(10):1664–1666

Verzijl D et al (2017) A novel label-free cell-based assay technology using biolayer interferometry. Biosens Bioelectron 87:388–395. https://doi.org/10.1016/j.bios.2016.08.095

Wade JL et al (1987) Theoretical description of nonlinear chromatography, with applications to physicochemical measurements in affinity chromatography and implications for preparative-scale separations. Anal Chem 59(9):1286–1295

Wang W et al (2012) Label-free measuring and mapping of binding kinetics of membrane proteins in single living cells. Nat Chem 4(10):846–853. https://doi.org/10.1038/NCHEM.1434

Wartchow CA et al (2011) Biosensor-based small molecule fragment screening with biolayer interferometry. J Comput Aided Mol Des 25(7):669–676. https://doi.org/10.1007/s10822-011-9439-8

Wei Z et al (2019) Specific TBC domain-containing proteins control the ER-Golgi-plasma membrane trafficking of GPCRs. Cell Rep 28(2):554-566.e4. https://doi.org/10.1016/j.celrep.2019.05.033

Wells CA et al (2012) Label-free detection of G protein-SNARE interactions and screening for small molecule modulators. ACS Chem Neurosci 3(1):69–78. https://doi.org/10.1021/cn200102d

White CW et al (2019) NanoBRET ligand binding at a GPCR under endogenous promotion facilitated by CRISPR/Cas9 genome editing. Cell Signal 54:27–34. https://doi.org/10.1016/j.cellsig.2018.11.018

Whited AM, Park PS (2014) Atomic force microscopy: a multifaceted tool to study membrane proteins and their interactions with ligands. Biochim Biophys Acta 1838(1PtA):56–68. https://doi.org/10.1016/j.bbamem.2013.04.011

Zeng KZ et al (2017) Rapid analysis of interaction between six drugs and β(2)-adrenergic receptor by injection amount-dependent method. Biomed Chromatogr 31(6):e3897. https://doi.org/10.1002/bmc.3897

Zhao XF et al (2014) Exploring drug-protein interactions using the relationship between injection volume and capacity factor. J Chromatogr A 1339:137–144. https://doi.org/10.1016/j.chroma.2014.03.017

Zhou H et al (2020a) Identification of novel phytocannabinoids from ganoderma by label-free dynamic mass redistribution assay. J Ethnopharmacol 246:112218. https://doi.org/10.1016/j.jep.2019.112218

Zhou XL et al (2020b) Surface plasmon resonance microscopy: from single-molecule sensing to single-cell imaging. Angew Chem Int Ed Engl 59(5):1776–1785. https://doi.org/10.1002/anie.201908806

Chapter 6
Immobilized GPCRs in Compound Screening

Abstract Screening the bioactive compounds targeting GPCRs is a pre-requisite step for the development of new drugs associated with the receptor. Synthetic compound libraries and natural product extracts are the widely used two sources for bioactive compound screening. In this chapter, we aim to review some representative immobilized GPCR-based methods for the discovery of the bioactive compounds targeting the receptors, which may provide insight into the high throughput screening of lead compounds.

Keywords Bioactive compound screening · Compound library · Natural products

Abbreviations

GPCRs	G protein-coupled receptors
CCR5	Chemokine receptor subtype 5
β_2-AR	Beta2-adrenoceptor
CXCR4	C-X-C motif chemokine receptor 4
$A_{2A}R$	Adenosine A_{2A} receptor
mGluR5	Metabotropic glutamate receptor 5
NTS_1R	Neurotensin receptor 1
D_1R	Dopamine receptor
M_2R	M2 muscarinic receptor
OR	Opioid receptor
CB_1R	Cannabinoid receptor 1
SPR	Surface plasmon resonance
DECLs	DNA-encoded chemical libraries
AT_1R	Angiotensin II type 1 receptor
ET_AR	Endothelin receptor A
HPAC	High-performance affinity chromatography
MS	Mass spectrometry
IAM	Immobilized artificial membrane
CMC	Cell membrane chromatography

X. Zhao et al., *G Protein-Coupled Receptors*, SpringerBriefs in Molecular Science,
https://doi.org/10.1007/978-981-99-0078-7_6

hAGT   Human O^6-alkylguanine-DNA alkyltransferase
HaloTag   Haloalkane dehalogenase tag
EGFR   Epidermal growth factor receptor
5-TH_{2C}R   Serotonin 5-hydroxytryptamine 2C receptor

6.1 Screening of Compound Libraries

The utilization of combinational chemistry for the pursuit of library synthesis has resulted in a steadily increasing number of new chemical entities (Abdelbary et al. 2019). Owing to this development, it becomes more and more popular to screen pooled compound libraries for the pursuit of drug candidates.

6.1.1 Sensor-Based Assays

Given the particular importance of the cost-effective identification of novel drug candidates, the screening of GPCR ligands from large compound libraries remains the major focus of drug discovery in both the infrastructure investment and the pharmaceutical industry. Among the competing techniques, the solid-phase plate-based methods are the most popular platform because they can be easily adapted for combination with other techniques. Typically, the plate-based methods are adapted for radioligand assays of entrapped GPCRs in terms of monolithic protein-doped materials that are fabricated in filter plates (Lebert et al. 2008). Alternatively, microwell-based radio assays can also be accomplished on membrane receptors by scintillation proximity assays. In this case, the GPCRs are immobilized on fluoro-microspheres containing the scintillant. The binding of radio-labeled ligands to the receptors produces a light signal that can be monitored (Harder and Fotiadis 2012). Such assays are advantageous since the radioisotope has little interference with binding behavior. However, the methods have not been broadly employed for drug discovery in the pharmaceutical industry given the difficulty of synthesizing radioactive ligands and their environmental issues.

　　To pursue label-free detection of ligand-receptor interaction, the development of new methods coupled with surface-sensitive detection has become one of the most interesting areas for plate-based assays. This has been realized by the utilization of gold-coated wells combined with surface plasmon resonance (Barbour and Bova 2012). Further improvement of the method has been performed by the use of a plate with a grating-based optical biosensor in each well that detects shifts in resonant wavelength upon ligand binding. The main limitation of these methods is nonspecific responses caused by temperature or nonspecific binding.

　　Owing to recent advances in GPCR expression and purification methods, the immobilization of purified GPCRs on SPR surfaces has emerged as a powerful strategy for screening compound libraries (Kumari et al. 2015). The first application of the method in GPCR ligand screening is performed on a thermo-stabilized adenosine A_{2A} receptor that yields several hits with affinities up to 10 μM (Congreve et al. 2011). Since then, similar strategies have been popularized for ligands screening of several GPCRs including chemokine receptor subtype 5 (CCR5) (Navratilova et al. 2011), human beta1-adrenoceptor (β_1-AR) (Christopher et al. 2013), C-X-C motif chemokine receptor 4 (CXCR4) (Martínez-Muñoz et al. 2015) and adenosine A_{2A} receptor (A_{2A}R) (Bocquet et al. 2015). These breakthroughs have highlighted the potential limitations of the method as the relatively low throughput and the stability issues of the screening protocol. Besides the advantages, these successful cases have also underlined the unique advantage of the method for reporting new ligand-binding pockets of the receptor. This leads to the screening of novel candidate ligands that are far different from the traditional allosteric or orthosteric ones.

Each of the above plate-based approaches displays its advantages and disadvantages when it is used for screening GPCR ligands from compound libraries. The ongoing application with high promise appears to be the hybrid use of them with other approaches like structure-based assays. The plate-based assays are powerful for initial ligand screening while structure-derived methods have priority of candidate optimization. Thus, their combination benefits the minimization of false-positive frequency. Application of such strategy has been reported recently in the ligand screening of metabotropic glutamate receptor 5 (mGluR5) where the combination of fragment screening and structure-guided optimization has generated a novel negative allosteric modulator of the receptor (Christopher et al. 2015). Taking advantage of this successful case, it is preferable to couple sensor-based assays with structure-based strategies, in particular when a few ligand candidates are identified.

6.1.2 GPCR Microarrays

The application of immobilized protein in microarray format appears as an interesting alternative for plate-based screening assays. This method has the merits of high throughput with the allowance of coupling to diverse detection assays such as colorimetry (Hamel-Côté et al. 2019), fluorescence (Zemella et al. 2019), and mass spectrometry (Clark et al. 2020). Relying on these distinct coupling utilizations, several assays can be devised to screen ligands from the compound library through protein–ligand interactions.

Since initial reports of protein microarray by Schreiber et al. (MacBeath and Schreiber 2000) and Zhu et al. (2001), enzyme microarrays have been well fabricated and applied in biomedical fields (Tu et al. 2014), which consists of the understanding of posttranslational modification and cell signaling pathways, and the determination of enzyme substrates and screening of drug candidates. However, it is not well achieved in GPCR fields, from the view of neither array fabrication nor the application in screening novel inhibitors, agonists, or antagonists with small molecular weight. Until now, only a few reports are surveyed as proof-of-concept studies. In this context, the first study by Fang et al. (2002) has accomplished the printing of the β_1-AR, β_2-AR, the neurotensin receptor 1 (NTS$_1$R), and the dopamine receptor (D$_1$R) on microarrays, which are subsequently applied in estimating the binding of compounds to the receptors. The subsequent report by Hong et al. (2005) has fabricated functional microarrays by immobilizing NTS$_1$R, M2 muscarinic receptor (M$_2$R), opioid receptor (OR), and cannabinoid receptor 1 (CB$_1$R) on porous glass slides, which are successful to monitor the activation of the receptors. Shortly afterward, the same group developed a microarray printing ten distinct GPCR fragments whereby the authors can measure the binding of fluorescently labeled ligands to the receptors (Hong et al. 2006). A recent work by Chadli et al. created a CXCR4 microarray by the combination of peptide-tethered bilayer lipid membrane formation and solid support micropatterning. The functionality of the immobilized GPCR was successfully analyzed by surface plasmon resonance (SPR) imaging and ligand binding studies (Chadli et al. 2018).

Despite these successes, GPCR microarrays for screening ligands from compound libraries have not yet been reported. It remains an emerging field even though the advantages of multiplexing and high amenability to other techniques. Such a situation may be caused by the infrastructure needed for the fabrication of the arrays which are not commonly available. In addition, current difficulties to purify high active GPCRs and following functionally immobilize them on the surface is likely the other factor that hinders the advance of GPCR microarrays in ligand screening. Once these difficulties are overcome, GPCR microarrays are expected to develop as a screening platform of choice.

6.1.3 Affinity Chromatography

In the column-based assays, GPCRs are immobilized on a suitable support and packed into a column. A compound library containing potential ligands is injected or continuously infused through the column in a "spike" or "breakthrough" form (Schriemer et al. 1998). Active ligands will bind to the column, which results in a longer retention time or breakthrough time than the non-retained ligands. To achieve a multi-dimensional analysis, the affinity columns are often interfaced with MS as the detector, where even a trace component is measurably distinct from other components at m/z value (Calleri et al. 2010). Compared to the sensor-based or microarrays, the column-based assays are the ideal method to examine an entire library to identify and rank the ligands in the library, and to ascertain the binding constants for each ligand in a single run.

In our previous work, we have introduced DNA encoding techniques into the synthesis of phenolic acid-focused libraries. This DNA-encoded chemical libraries (DECLs) contain 32,000 diverse compounds. We first obtained two potential compounds that target angiotensin II type 1 receptor (AT$_1$R) and could dramatically reduce the blood pressure of renovascular hypertensive rats (Liang et al. 2021a). Using the same DECLs, we also screened the bioactive compounds targeting endothelin receptor A ($ET_A R$) (Liang et al. 2021b). The potential allosteric modulator of β_2-AR was also found from the natural product derived DECLs (Tian et al. 2022). Another work applied rosmarinic acid, 18b-glycyrrhetinic acid, rhein, and ferulic acid as template building blocks to construct DECLs containing 1,000 compounds. Liang et al. found a dual-target ligands binding to β_2-AR and cysteinyl-leukotriene receptor for the potential treatment of asthma (Liang et al. 2022). These works offered insight into the affinity chromatography in high throughput screening of bioactive compounds from DECLs.

6.2 Screening of Natural Product Extracts

Natural product extracts have diverse chemical structures and distinct bioactivities, which have proved to be effective for the treatment of diseases during the past thousand years. Different from the compound library, natural products avoid sophisticated synthesis process, thus, being a more attractive source in drug discovery. The components in natural products are often more complicated than in the compound library, bringing a lot of challenges using the sensor based and microarray methods.

Due to the high separation capacity of high-performance affinity chromatography (HPAC), its online hyphenation to mass spectrometry (MS) becomes the most popular way in the screening of natural product extracts. Pioneering work of I. W. Wainer and R. Moaddel immobilized GPCRs, including ORs (Beigi and Wainer 2003), β_2-AR (Beigi et al. 2004), nicotinic acetylcholine receptors (Baynham et al. 2002), and purinergic receptor (Moaddel et al. 2007a, b), on the immobilized artificial membrane (IAM) silica gels to prepare the affinity chromatographic stationary phases. They used these stationary phases for drug-receptor interaction analysis and online screening. For instance, they screened the bioactive components that target the nicotinic acetylcholine receptor for smoking cessation from the extract of *Lycopodium clavatum*, *Trigonella foenum graecum*, and tobacco smoke condensate (Ciesla et al. 2016; Maciuk et al. 2008). On the basis of these successful application, their group immobilized cell membrane fragments with cannabinoid receptors onto open tubular capillary surface and further identified the targeted ligands from the extract of *Zanthoxylum bungeanum* (Dossou et al. 2013). In 1990s, He et al. proposed a biomimetic chromatographic method called cell membrane chromatography (CMC) (Hou et al. 2014). They immobilized more than twenty kinds of cell membranes containing targeted GPCRs to support surfaces as affinity stationary phases and applied them in natural product screening. During the last two decades, they found about 60 active components from natural products, including 25 antineoplastics, 12 anti-cardiovascular compounds, 16 antiprostatic hyperplasia drugs, and 8 compounds for other activities (Ma et al. 2021). Although the CMC and IAM models are valid for drug screening, a series of work are still needed to demonstrate that the obtained bioactive compounds are indeed targeting the GPCRs, but not to the cell or artificial membranes.

To address this issue, our group immobilized purified GPCRs on the surface of solid supports and proposed a new type of affinity chromatographic method, namely, receptor chromatography (Zhao et al. 2012; Jia et al. 2019). Due to the lack of cell membrane around the immobilized GPCRs, the stability of the receptors become a major challenge. During the past ten years, we have developed a series of self-labeling active enzyme tags for GPCR immobilization, including Human O^6-alkylguanine-DNA alkyltransferase (hAGT), haloalkane dehalogenase tag (HaloTag), and epidermal growth factor receptor (EGFR tag), and dedicated to improve the stability and activity of the immobilized GPCRs (Wang et al. 2019; Zeng et al. 2018; Zhao et al. 2020). In addition, we used RNA and DNA aptamer to stabilize the GPCR conformation and the stabilized GPCRs were successfully applied in bioactive compound screening from traditional Chinese medicine (Gao et al. 2020; Gao et al. 2021; Liu et al. 2022).

Another plausible approach is to use the protein coated microspheres or magnetic beads for natural product screening, namely, ligand fishing (Moaddel et al. 2007a, b; Marszałł et al. 2008). In this method, when the beads are immersed into the extract: (i) non-binders remain in the supernatant and can be separated with the beads through centrifugation or magnetic force; (ii) the compounds that have affinity to the immobilized proteins are retained on the beads and can be eluted by the buffer with strong elution capacity or protein denaturation reagent. As pioneers for ligand fishing, R. Moaddel et al. has immobilized serum protein (human serum albumin) and enzymes (tyrosine phosphatase, a-glucosidase, sirtuin 6, acetylcholinesterase, and neuraminidase) on the magnetic nanoparticles and fished out many ligands, especially the low-affinity ligands, towards the target proteins from natural product extract (Cieśla and Moaddel 2016; Wubshet et al. 2015; Zhao et al. 2018; Yasuda et al. 2011). Our group extended their work in GPCR targeted ligand fishing from traditional Chinese medicine compound formulations. We used the β_2-AR coated silica gels and magnetic beads for ligand fishing from Xie-Bai-San and San-Ao decoction, respectively, and successfully screened eight compounds that target the receptor (Fei et al. 2018; Sun et al. 2017). In a recent work, Shui's group used serotonin 5-hydroxytryptamine 2C receptor (5-TH$_{2C}$R) coated nickel agarose beads to fish the bioactive compounds in the extracts of *Stephania tetrandra*, *Aristolochia debilis* and *Tetradium ruticarpum*. They identified a novel subtype selective agonist (*R*)-asimilobine (1857) for 5-TH$_{2C}$R, which showed comparable anti-obesity effect as lorcaserin and exhibited exclusive bias towards G protein signaling (Zhang et al. 2020).

With the increasing government restrictions on drug approvals, it is more challenging to find effective drugs directly from synthetic libraries and natural product extract. Structural modification is inevitable to achieve the optimal drug efficiency and minimal toxicity. The recent development of machine learning approaches provides a set of tools that can accelerate discovery (Vamathevan et al. 2019). Despite this, screening the high affinity ligands with desirable drug-like property is still vital to reduce the high unpredictability.

References

Abdelbary HM et al (2019) Characterization and radiological impacts assessment of scale TENORM waste produced from oil and natural gas production in Egypt. Environ Sci Pollut Res Int 26:30836–30846. https://doi.org/10.1007/s11356-019-06183-x

Barbour R, Bova MP (2012) Combining label-free technologies: discovery in strength. Bioanalysis 4:619–622. https://doi.org/10.4155/bio.12.45

Baynham MT et al (2002) Multidimensional on-line screening for ligands to the $\alpha3\beta4$ neuronal nicotinic acetylcholine receptor using an immobilized nicotinic receptor liquid chromatographic stationary phase. J Chromatogr B Analyt Technol Biomed Life Sci 772:155–161. https://doi.org/10.1016/S1570-0232(02)00070-3

Beigi F et al (2004) G-protein-coupled receptor chromatographic stationary phases. 2. Ligand-induced conformational mobility in an immobilized β_2-adrenergic receptor. Anal Chem 76:7187–7193. https://doi.org/10.1021/ac048910c

Beigi F, Wainer IW (2003) Syntheses of immobilized G protein-coupled receptor chromatographic stationary phases: characterization of immobilized mu and kappa opioid receptors. Anal Chem 75:4480–4485. https://doi.org/10.1021/ac034385q

Bocquet N et al (2015) Real-time monitoring of binding events on a thermostabilized human A2A receptor embedded in a lipid bilayer by surface plasmon resonance. Biochim Biophys Acta 1848:1224–1233. https://doi.org/10.1016/j.bbamem.2015.02.014

Calleri E et al (2010) Frontal affinity chromatography-mass spectrometry useful for characterization of new ligands for GPR17 receptor. J Med Chem 53:3489–3501. https://doi.org/10.1021/jm901691y

Chadli M et al (2018) A new functional membrane protein microarray based on tethered phospholipid bilayers. Analyst 143:2165–2173. https://doi.org/10.1039/C8AN00260F

Christopher JA et al (2013) Biophysical fragment screening of the β_1-adrenergic receptor: identification of high affinity arylpiperazine leads using structure-based drug design. J Med Chem 56:3446–3455. https://doi.org/10.1021/jm400140q

Christopher JA et al (2015) Fragment and structure-based drug discovery for a class C GPCR: discovery of the mGlu5 negative allosteric modulator HTL14242 (3-chloro-5-[6-(5-fluoropyridin-2-yl)pyrimidin-4-yl]benzonitrile). J Med Chem 58:6653–6664. https://doi.org/10.1021/acs.jmedchem.5b00892

Ciesla L et al (2016) Development and characterization of the $\alpha3\beta4\alpha5$ nicotinic receptor cellular membrane affinity chromatography column and its application for on line screening of plant extracts. J Chromatogr A 1431:138–144. https://doi.org/10.1016/j.chroma.2015.12.065

Cieśla Ł, Moaddel R (2016) Comparison of analytical techniques for the identification of bioactive compounds from natural products. Nat Prod Rep 33:1131–1145. https://doi.org/10.1039/c6np00016a

Clark LJ et al (2020) Allosteric interactions in the parathyroid hormone GPCR-arrestin complex formation. Nat Chem Biol 16:1096–1104. https://doi.org/10.1038/s41589-020-0567-0

Congreve M et al (2011) Fragment screening of stabilized G-protein-coupled receptors using biophysical methods. Methods Enzymol 493:115–136. https://doi.org/10.1016/B978-0-12-381274-2.00005-4

Dossou KS et al (2013) Identification of CB1/CB2 ligands from *Zanthoxylum bungeanum*. J Nat Prod 76:2060–2064. https://doi.org/10.1021/np400478c

Fang Y, Frutos AG, Lahiri J (2002) Membrane protein microarrays. J Am Chem Soc 124:2394–2395. https://doi.org/10.1021/ja017346+

Fei F et al (2018) Rapid screening and identification of bioactive compounds specifically binding to beta2-adrenoceptor from San-ao decoction using affinity magnetic fine particles coupled with high-performance liquid chromatography-mass spectrometry. Chin Med 13:49. https://doi.org/10.1186/s13020-018-0207-8

Gao J et al (2020) Reversible and site-specific immobilization of β₂-adrenergic receptor by aptamer-directed method for receptor-drug interaction analysis. J Chromatogr A 1622:461091. https://doi.org/10.1016/j.chroma.2020.461091

Gao J et al (2021) Two-point immobilization of a conformation-specific beta2-adrenoceptor for recognizing the receptor agonists or antagonists inspired by binding-induced DNA assembly. Biomater Sci 9:7934–7943. https://doi.org/10.1039/d1bm01222c

Hamel-Côté G, Lapointe F, Stankova J (2019) Measuring GPCR-induced activation of protein tyrosine phosphatases (PTP) using In-gel and colorimetric PTP assays. In: Tiberi M (eds) G protein-coupled receptor signaling. Methods in molecular biology, vol 1947. Humana Press, New York, NY. https://doi.org/10.1007/978-1-4939-9121-1_13

Harder D, Fotiadis D (2012) Measuring substrate binding and affinity of purified membrane transport proteins using the scintillation proximity assay. Nat Protoc 7:1569–1578. https://doi.org/10.1038/nprot.2012.090

Hong Y et al (2005) Functional GPCR microarrays. J Am Chem Soc 127:15350–15351. https://doi.org/10.1021/ja055101h

Hong Y et al (2006) G-protein-coupled receptor microarrays for multiplexed compound screening. J Biomol Screen 11:435–438. https://doi.org/10.1177/1087057106287139

Hou XF et al (2014) Recent advances in cell membrane chromatography for traditional Chinese medicines analysis. J Pharm Biomed Anal 101:141–150. https://doi.org/10.1016/j.jpba.2014.05.021

Jia X et al (2019) Screening bioactive compounds of *Siraitia grosvenorii* by immobilized β₂-adrenergic receptor chromatography and druggability evaluation. Front Pharmacol 10:915. https://doi.org/10.3389/fphar.2019.00915

Kumari P, Ghosh E, Shukla AK (2015) Emerging approaches to GPCR ligand screening for drug discovery. Trends Mol Med 21:687–701. https://doi.org/10.1016/j.molmed.2015.09.002

Lebert JM, Forsberg EM, Brennan JD (2008) Solid-phase assays for small molecule screening using sol-gel entrapped proteins. Biochem Cell Biol 86:100–110. https://doi.org/10.1139/O08-010

Liang Q et al (2021a) Selective discovery of GPCR ligands within DNA-encoded chemical libraries derived from natural products: a case study on antagonists of angiotensin II type I receptor. J Med Chem 64:4196–4205. https://doi.org/10.1021/acs.jmedchem.1c00123

Liang Q et al (2021b) Identification of selective ligands targeting two GPCRs by receptor-affinity chromatography coupled with high-throughput sequencing techniques. Bioorg Chem 112:104986. https://doi.org/10.1016/j.bioorg.2021.104986

Liang Q et al (2022) Discovery of dual-target ligands binding to beta2-adrenoceptor and cysteinyl-leukotriene receptor for the potential treatment of asthma from natural products derived DNA-encoded library. Eur J Med Chem 233:114212. https://doi.org/10.1016/j.ejmech.2022.114212

Liu JJ et al (2022) Aptamer-assisted two-point immobilized agonist-bound angiotensin II type 1 receptor for a second site modulator discovery. iScience 25:105361. https://doi.org/10.1016/j.isci.2022.105361

Ma W et al (2021) Advances in cell membrane chromatography. J Chromatogr A 639:461916. https://doi.org/10.1021/np400478c

MacBeath G, Schreiber SL (2000) Printing proteins as microarrays for high-throughput function determination. Science 289:1760–1763. https://doi.org/10.1126/science.289.5485.1760

Maciuk A et al (2008) Screening of tobacco smoke condensate for nicotinic acetylcholine receptor ligands using cellular membrane affinity chromatography columns and missing peak chromatography. J Pharm Biomed Anal 48:238–246. https://doi.org/10.1016/j.jpba.2007.11.024

Marszałł MP et al (2008) Ligand and protein fishing with heat shock protein 90 coated magnetic beads. Anal Chem 80:7571–7575. https://doi.org/10.1021/ac801153h

Martínez-Muñoz L et al (2015) Methods to immobilize GPCR on the surface of SPR sensors. In: Prazeres DMF, Martins SAM (eds) G protein-coupled receptor screening assays. Methods in molecular biology, vol 1272. Humana, New York, NY. https://doi.org/10.1007/978-1-4939-2336-6_12

Moaddel R et al (2007a) Automated ligand fishing using human serum albumin-coated magnetic beads. Anal Chem 79:5414–5417. https://doi.org/10.1021/ac070268+

Moaddel R et al (2007b) The synthesis and initial characterization of an immobilized purinergic receptor (P2Y1) liquid chromatography stationary phase for online screening. Anal Biochem 364:216–218. https://doi.org/10.1016/j.ab.2007.02.014

Navratilova I, Besnard J, Hopkins AL (2011) Screening for GPCR ligands using surface plasmon resonance. ACS Med Chem Lett 2:549–554. https://doi.org/10.1021/ml2000017

Schriemer DC et al (1998) Micro-scale frontal affinity chromatography with mass spectrometric detection: a new method for the screening of compound libraries. Angew Chem Int Ed Engl 37:3383–3387. https://doi.org/10.1002/(SICI)1521-3773(19981231)37:24%3c3383::AID-ANIE3383%3e3.0.CO;2-C

Sun Z et al (2017) A fast affinity extraction methodology for rapid screening of bioactive compounds specifically binding to beta2-adrenergic receptor from *Xie-Bai-San*. Med Chem Res 26:2410–2419. https://doi.org/10.1007/s00044-017-1941-7

Tian R et al (2022) Development of an allostery responsive chromatographic method for screening potential allosteric modulator of beta2-adrenoceptor from a natural product-derived DNA-encoded chemical library. Anal Chem 94:9048–9057. https://doi.org/10.1021/acs.analchem.2c01210

Tu S et al (2014) Protein microarrays for studies of drug mechanisms and biomarker discovery in the era of systems biology. Curr Pharm Des 20:49–55. https://doi.org/10.2174/138161282001140113123707

Vamathevan J et al (2019) Applications of machine learning in drug discovery and development. Nat Rev Drug Discov 18:463–477. https://doi.org/10.1038/s41573-019-0024-5

Wang J et al (2019) Site-specific immobilization of β_2-AR using O_6-benzylguanine derivative functionalized supporter for high-throughput receptor-targeting lead discovery. Anal Chem 91:7385–7393. https://doi.org/10.1021/acs.analchem.9b01268

Wubshet SG et al (2015) Magnetic ligand fishing as a targeting tool for HPLC-HRMS-SPE-NMR: α-glucosidase inhibitory ligands and alkylresorcinol glycosides from *Eugenia catharinae*. J Nat Prod 78:2657–2665. https://doi.org/10.1021/acs.jnatprod.5b00603

Yasuda M et al (2011) Synthesis and characterization of SIRT6 protein coated magnetic beads: identification of a novel inhibitor of SIRT6 deacetylase from medicinal plant extracts. Anal Chem 83:7400–7407. https://doi.org/10.1021/ac201403y

Zemella A et al (2019) A combined cell-free protein synthesis and fluorescence-based approach to investigate GPCR binding properties. In: Tiberi M (ed) G protein-coupled receptor signaling. Methods in molecular biology, vol 1947. Humana Press, New York, NY. https://doi.org/10.1007/978-1-4939-9121-1_4

Zeng K et al (2018) One-step methodology for the direct covalent capture of GPCRs from complex matrices onto solid surfaces based on the bioorthogonal reaction between haloalkane dehalogenase and chloroalkanes. Chem Sci 9:446–456. https://doi.org/10.1039/c7sc03887a

Zhang B et al (2020) A novel G protein-biased and subtype-selective agonist for a G protein-coupled receptor discovered from screening herbal extracts. ACS Cent Sci 6:213–225. https://doi.org/10.1021/acscentsci.9b01125

Zhao X et al (2012) Using immobilized G-protein coupled receptors to screen bioactive traditional Chinese medicine compounds with multiple targets. J Pharm Biomed Anal 70:549–552. https://doi.org/10.1016/j.jpba.2012.05.004

Zhao YM et al (2018) Magnetic beads-based neuraminidase enzyme microreactor as a drug discovery tool for screening inhibitors from compound libraries and fishing ligands from natural products. J Chromatogr A 1568:123–130. https://doi.org/10.1016/j.chroma.2018.07.031

Zhao XF et al (2020) Covalent inhibitor-based one-step method for endothelin receptor A immobilization: from ligand recognition to lead identification. 92:13750–13758. https://doi.org/10.1021/acs.analchem.0c01807

Zhu H et al (2001) Global analysis of protein activities using proteome chips. Science 293:2101–2105. https://doi.org/10.1126/science.1062191